THE WORLD'S GREAT

AUTOMOBILE STYLISTS

THE WORLD'S GREAT

AUTOMOBILE STYLISTS

JOHN TIPLER

MALLARD
PRESS

MALLARD PRESS

An Imprint of BDD Promotional Book Company, Inc.
666 Fifth Avenue
New York, N.Y. 10103

Mallard Press and its accompanying design and logo are
trademarks of BDD Promotional Book Company, Inc.

First published in the United States of America by
The Mallard Press

ISBN 0-792-45345-X

This book was designed and produced by
Quintet Publishing Limited
6 Blundell Street
London N7 9BH

Art Director: Peter Bridgewater
Designer: Annie Moss
Editor: Shaun Barrington

Typeset in Great Britain by
Central Southern Typesetters, Eastbourne
Manufactured in Hong Kong by
Regent Publishing Services Limited
Printed in Hong Kong by
Leefung Asco Printers Limited

ACKNOWLEDGEMENTS

I should like to thank everyone who assisted
me in preparing the World's Great Automobile
Stylists, especially the National Motor Museum
at Beaulieu, and Dr Paul Nieuwenhuis of the
Motor Industry Research Unit, Norwich. The
book is dedicated to my new daughter Zoë
who very kindly hung on until I'd finished
writing it before she asked to be born!

John Tipler

CONTENTS

INTRODUCTION

THE JAZZ AGE
THE GRAND ROUTIERI
POP CULTURE

Who are these mystery men who create the shapes of the cars we drool over at motor shows, the cars we'd long to have standing in our own driveways? What are their backgrounds? Generally speaking, they're enthusiasts first, draughtsmen second, design engineers third. That seems to be the basic makeup of the auto stylist. To produce great designs, you have to be passionately addicted to your subject, and of course, you have to have a flare with the pen. But because you are merely designing the car's clothes, as it were, it's not necessarily vital to have more than a sound grasp of what goes on under the hood.

In the early days, the preoccupation was with cladding the mechanicals in some sort of practical fashion, so that the electrics didn't get soaked, nor indeed did the occupants. Coach-builders, as the name implies, naturally knew more about the craft of building horse-drawn vehicles, and the closed carriage was almost unknown in the first two decades of the motor car. Right up until the 1950s, particularly in the luxury market, it was very much the regular thing to buy a rolling chassis from the manufacturer and have a specialist coachbuilder fit one of his off-the-shelf bodies. Sometimes the coachbuilder was also the stylist, or had a stylist working for him, but generally, he was an artisan, a fabricator, rather than a stylist working on a drawing board and with a clay model.

Because there were so few motor cars about, they were inevitably different stylistically, but really not in any conscious way. Broken down into its fundamentals, the car consists of four wheels, an engine, somewhere for its driver to sit and a receptacle for its fuel. Styling is merely a matter of how you deal with the dressing-up of these fundamentals. Probably the first car to feature distinctive styling was the Mercer Race-about of 1908. Its designer was Finlay Robertson Porter, and it set the standard for a generation of sporting cars, with sweeping mudguards and running-boards from front to back, twin seats with fuel tank behind, and a neat but functional cover over the bonnet. So did many other cars of the period, but the 75mph Mercer has that extra panache, that elemental austerity which is great styling.

While the Mercer was being created, Henry Ford was creating the motor industry with his Model-T; and earlier, in 1912, William Morris launched the Oxford, a car about as exciting to look at as the Model-T, but innovative insofar as it was assembled from bought-in components. It set the scene for a generation of passenger cars. Andre Citroen took matters a stage further with his Model A of 1919 when he imported most of the plant to Paris from the USA.

But to a large extent it is at the sporting end of the vehicular spectrum where the great stylists lavish their talents. This is because speed

BELOW Major manufacturers reserve their most advanced styling for sports models. Seen here is Honda's NSX, a typical example for the 1990s.
BOTTOM Various models of prototypes on display at the Henry Ford Museum, Dearborn.

■ The sparsely-clad Mercer Raceabout of 1910 was typical of a generation of veteran sports cars.

has always had a romantic fascination, and can be further glamorized by extravagant styling. Sports car builders invariably operate on relatively short production runs, and this is true of major manufacturers who offer models aimed at the sports market. They can call upon a well-known stylist or styling house to give their sports model a distinctive look.

If the Mercer represented the sporting side of the market in 1910, the Hispano Suiza Alfonso of 1913 was the flagship of the slightly more staid Grand Tourers. Again, although the mechanical engineering of the Hispano Suiza was designed by Marc Birkigt, it was clad in a variety of bodies from a number of coachbuilders. One of the nicest was done in 1912 by Lookers coachworks in Manchester. A one-off sports model was even styled in 1913 by the King of Spain, Alfonso XIII. Contemporary with the Hispano were a few other notable sporting cars; the 1914 Mercedes Benz, Isotta Fraschini, Marion, Fiat, Nazzaro, Itala, National, Prince Henry Vauxhall, and Peugeot. All these makes achieved conspicuous racing successes, but elaborate styling exercises were virtually unknown. Perhaps the Pierce-Arrow designed by George N. Pierce in 1913 was the first in this field, with its fared-in headlamps integral with the mudguards, deliberately setting out to obtain a distinctive shape.

In a way, Ford and Morris were ahead of their time, for western society at the turn of the century was in the grip of the romantically decadent Art Nouveau, the rustic Arts and Crafts movement, and the fripperies of late Victoriana. The art world was being turned upside down by the Cubists and other modernist groups like the Futurists, the latter venerating the motor car as the symbol of the dynamic new age. Fiat's Lingotto factory near Turin with its rooftop banked-bend racetrack was designed by Matte Trusco, a Futurist architect. Not surprisingly, the evolution of the automobile body style was checked until after World War I. One of the main effects of the Great War was the establishment of production lines to meet the insatiable demand for basic weaponry and munitions. Motor manufacturers like Rolls Royce and Hispano Suiza provided their engines for aircraft and military vehicles. The war produced few significant technological advances, with turbocharging in aircraft a notable exception; industrially, it was all about building a great many identical machines and components. But if the pre-war decade and the period of hostilities had yielded little in terms of aesthetics, the 1920s saw vehicle styling really begin to take off, to the point where it became an integral part of marketing. It was the racing machines which led the way, with their quest for streamlined shapes constantly pushing back the speed frontier.

Foremost innovator was Ettore Bugatti, whose 1923 'tank' bodied grand prix cars were clad in all-enveloping bodywork which had an aerofoil

Mass production saw the beginnings of organised styling. The Fiat 501 of 1919 is still concerned with cladding mechanicals and passengers. By 1930, Packard's grand design reflected the status of the owner.

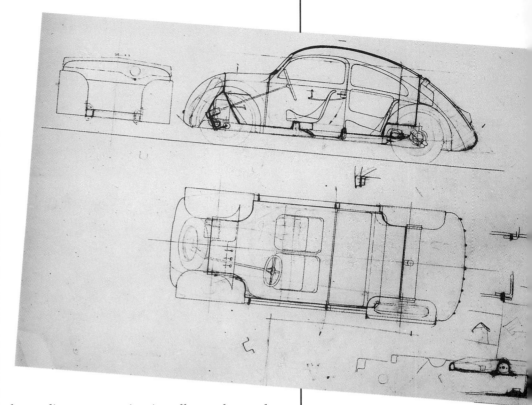

configuration. Whereas today we know all about reverse lift and downforce generated by inverted aerofoils, Bugatti didn't, and the Type 30 body, shaped in cross-section like the wing of an aircraft, was decidedly unstable around its 120mph top speed. In addition, it was thought publicly to be unattractive, so Bugatti abandoned it in order to preserve a full order book. The pretty, low-slung Type 35 which followed in 1924 was also innovative as the first car to have cast alloy wheels with integral brake drums. Bugatti's chief rivals at the time were Ballot, Sunbeam and Fiat, and of course Bentley, who was following a very different course. The Bentley sports-racing cars of the 1920s, described by Bugatti as the 'fastest trucks in the world', were brutish and archaic in comparison with the light and attractive French creation. Yet on the race track, they were hugely successful. More brutal still was Mercedes-Benz's SSLK, designed by Ferdinand Porsche in 1929. Less aggressive were models from Lagonda and Vauxhall, whose 30/98 preceded the company's General Motors takeover and subsequent subordination to Harley Earl. Elsewhere in the USA, coachbuilders Brewsters were making, according to some, a better job of cladding US-built Rolls Royces than the generally dull bodies produced in the UK.

THE JAZZ AGE

The Bauhaus, Le Corbusier, Art Deco and Der Stijl in Europe, Jazz and Hollywood in the US: but the automobile stylists lagged dreadfully behind in the modernist explosion, paying mere lip service to Art Deco only in muted dashboard styling, motifs and radiator mascots. The Bugatti Type 35 and the Alfa Romeo 6C 1750 of 1929 exemplify the best in car styling during the 1920s.

Between the First and Second World Wars, the French stylists were dominant, although a number of outstanding designs emanated from US drawing boards. The rich and fashionable were traversing the globe, and in Europe in particular it was very much the thing to do to motor down to the Côte d'Azure, the French Riviera. To do this in style, the 1930s fast set turned to the sporting models of the day, and very soon these 'Grand Routieri' were to be seen speeding southward to the sea and the sunshine. Getting on the road with your new grand tourer was actually not quite as straightforward as that. Generally, the customer ordered his chosen chassis and power unit, and a design was selected from the stylist. The coachbuilder then made a model in wood, and perhaps even a full-scale mockup in order to give a good idea of what the finished article would look like. When all was approved, the vehicle would be assembled, and off you went. In broad terms,

the styling process is virtually unchanged today, except for different mediums for the model, like clay (intoduced in the late 1920s by Harley Earl), or plaster, plasticine, and glass-reinforced-plastic (GRP); and the advent of wind-tunnels and computer-aided-design (CAD), to develop the designer's original ideas.

THE GRAND ROUTIERI

A number of excellent French coachbuilders came to prominence during the 1930s: Saoutchik, Faget Varnet, Franay, Pourtout, Figoni et Falaschi, all producing the most delightfully elegant and sometimes extravagant car bodies. The chief French chassis were provided by Bugatti, Delage, Delahaye, Hotchkiss, Salmson, and Talbot. Their appeal was strictly limited to

■ Dr Ferdinand Porsche brought streamlining to popular cars with his Typ-60, the prototype Volkswagen. Citroen's Light 15 with its advanced engineering concepts spelt the end for the Grand Routieri.

Andre Citroen's 7CV coupe of 1938 was ahead of most of the competition both stylistically and mechanically.

the wealthiest market. The cars tended to be heavy and uneconomical because of their overly heavy chassis and bodywork. Two cars spelled the end for the carrossiers of the Grand Routieri: the first was the light, monocoque front-wheel-drive Citroen 7 (later to become the Light 15), and the second the contemporary Peugeot, which had excellent road-holding and offered economical and comfortable motoring in considerable style, even in 1934. The Citroen, above all, showed the writing was on the wall for the luxurious leviathans, but the end really came with World War II and the occupation of France. After hostilities ceased, many of the coachbuilders tried to get going again with their pre-

war designs, but, of course, times had changed: not only in the automobile market, but also in the availability of raw materials. Some tried using wood, making perhaps as much as three quarters of the body in wood and the rest in steel. A Daimler rebodied in this fashion by Saoutchik in 1949 ended up weighing a ton more than it did with its factory body!

The French coachbuilders appeared not to have grasped that the market and indeed the whole trend of automotive design in Europe had changed, and their fate was more or less sealed by the swingeing tax on luxury cars in 1950s France. Mass-market designs in Europe, like Standard's Vanguard, were heavily influenced by the USA, although in the specialized sector, the mantle was swiftly taken up by the Italians like Ghia and Pinin Farina. The war had fostered major developments in aerodynamics, mechanical and metallurgic technology, and it seemed as if a pent-up wave of creativity was about to break in the buoyant atmosphere of liberated Italy. It didn't happen in Britain, doomed to a decade of stylistic mediocrity thanks to an economy devastated by five years of war. Many of Britain's factories may have survived, but industry was moribund, and in a climate of enforced austerity, functionalism abounded, from Utility furniture to Land Rover; the one bright spot was William Lyons' Jaguar XK120. Even the small French manufacturers forsook their native carrosserie for Italy, and certainly by 1950, the big manufacturers like Peugeot and Renault were aware of the market-

ing advantages of being able to identify a prominent Italian stylist as having penned one of their models. It was said in the early 1950s that one automatically went to Italy for one's car 'body', just as one went to visit a tailor in London! And this despite the fact that in his early days, Pinin Farina turned out only three or four designs a month. His Lancia Aurelia B20, launched at the 1951 Turin Show and in production from 1953, has a softness of line which heralds a new sense of proportion for the 1950s, in Europe at any rate. First of the new-look sporting machines was another Pinin Farina effort, the alloy-bodied Cisitalia 202, a light fresh design which influenced a further generation of coupés.

Rational styling in the 1950s, where aesthetics were paramount, belonged to the Italian styling houses of Pinin Farina, Touring and Ghia. In the US, things became increasingly outrageous, due to the obsessions of the principal stylists – Harley Earl, Bill Mitchell, Dutch Darrin and Raymond Loewy – first with jet fighters and then space rockets. In the mass market, stylistic changes were done as a matter of course as a means of selling more cars. In Britain, the market followed the US and Italy in diluted form, while there was a more realistic attitude toward aerodynamics in the Frank Costin and Colin Chapman-designed Lotus sports-racing and Vanwall grand prix cars. British laurels in international sports car racing were won by Jaguar

■
OPPOSITE Front and rear views of the 1959 Cadillac at GM's proving ground.
BOTTOM Long-wheelbase version of the Citroen Traction Avant. **LEFT** Jaguar XK 120 was pride of Britain in 1954.
BELOW Cisitalia 202 set new standards for sports coupe styling.

Contrasts in convertibles: **RIGHT** straight off the Detroit drawing board in 1956, an idea for a new Ford convertible. **BELOW** The European way; Citroens have always been ahead on style, but Henri Chapron turned this DS into a convertible.

with its curvaceous C- and D-Type cars, followed by Aston Martin's DB-Rs. Like the Mercedes which flowered briefly in mid-decade, these were home grown, if purposefully attractive, versions of earlier Italian-designed Ferraris, Maseratis, and Alfa Romeos like Alfredo Vignale's 1950 Carrozzeria Sport.

POP CULTURE

The great British conceptual breakthrough came in 1959 with Alex Issigonis' Mini, which fitted in with other elements of 'trendy' design exemplifying the pop culture. The popular climate revolved around youth culture, and there was a significant wealth explosion, providing greater disposable income for car buying. In terms of vehicle design, the Mini fostered a new generation of small front-wheel-drive town cars, and popularized the front-drive layout which Citroen

had taken pretty much for granted as the normal thing for twenty years. At the glamorous end of the market, it was again the Pininfarina Ferraris, Malcolm Sayer's E-Type Jaguar, and, slightly down market from those, the Lotus Elan, which epitomized desirable sports styling. It was a time of extravagant colour schemes, directly influenced by the fashion houses of swinging London. In the US, the 'pony' cars took off with the Ford Mustang in 1964, and the Corvette got steadily meaner-looking. As the decade sprinted on, supercars like the Bertone studio's Lamborghini Miura and the Eric Broadley-derived Ford GT40 captured the imagination. It was sports-racers like the GT40, Ferrari 250LM and P3/4, plus Broadley's own Lola T70, and the Porsche 906, which set the standard for sports racing cars. Porsche's own succession of 906, 907, 908 and 917 is a clear example of how such a design progresses, traceable right down to the 962s racing in 1989. During the 1960s and 1970s, tyre technology progressed through competition development to such an extent that the wide tyres seen on the racers became a regular feature of the road-going supercar. Anatole Lapine's Porsche 928 is a case in point, while Ferdinand 'Buzzi' Porsche's 911, conceived in 1963, now sports wheels so wide they would have looked positively bizarre 25 years ago.

A great many anonymous bread-and-butter designs are 'styled-by-committee', (the Ford Capri is a good example), insofar as there is

LEFT AND BELOW The Lamborghini Muira took its inspiration from sports racing cars of the mid '60s. *BELOW LEFT* The V8 engined 928 was Porsche's Grand Tourer for the 1970s.

LEFT AND OPPOSITE Styled by committee, Ford's Capri, the "car you always promised yourself" takes shape in 1968. Despite old-fashioned mechanicals, it remained popular for two decades.

input from a number of draughtsmen working in the design department, and the top-brass also has its say. However, men like Earl, Loewy and Mitchell were virtually household names in the US, and by the 1970s, the same was starting to become true in Europe. Stylists like Giugiaro and Gandini had served their time with the old houses of Ghia and Bertone, and were busy setting up studios of their own. Giugiaro's Ital Design consultancy was chiefly responsible for the spate of angular, wedge-shaped cars, so popular with the British in the early 1970s. Best examples of the 'wedge' are the Lotus Excel and second-generation Elite. A stunning show-car every year doesn't pay the rent, even if it does enhance the showcase, and these men followed in the footsteps of their peers by producing styles for major manufacturers. There ensued a progression of logical but unexciting

ABOVE A wooden mock-up for the 'jellymould' Sierra takes shape in the studios at Ford.

RIGHT Giugiaro's study for a luxury Lotus, the Etna.

BELOW The Ferrari F40 of 1988 makes no secret of its competition heritage.

BOTTOM Competition Technology was affordable in the Lotus Elan S1.

'Eurobox' hatchbacks, which again came to embrace competition aerodynamic aids in order to make them look interesting, so that city-centre chic demanded spoilers, air dams and side skirts on cars which could barely exceed 100mph. This fashion had been rationalized by 1988 so that all these excrescences had been incorporated into the basic design of a car like the Fiat Tipo. Cars like Uwe Bahnsen's 'jellymould' Ford Sierra took interesting styling into the popular market, and the uncompromising Audi Quattro popularized turbocharging and four-wheel-drive. Pininfarina's Peugeot 205, the Tipo and its older but smaller sister the Uno, were popular 'designer' cars, in so far as they were crisp, rationalized shapes, and the newly-wealthy 'yuppie' culture of the capital cities embraced clean new designs like the BMW Z1, and the two greatest supercars, the Porsche 959 and the Ferrari F40.

Many manufacturers had been employing the wind tunnel to improve vehicle aerodynamics for a decade and a half before these models were introduced; Henry deSegur Lauve's Citroen-Maserati SM achieved a cd figure as

low as 0.25 in the late 1960s. However, a new influence was waiting in the wings, and by 1980, the computer was a sufficiently sophisticated piece of equipment to be a viable tool in the design studio. Computer-aided-design soon went hand-in-hand with computer-aided-manufacturing, a practice pioneered by the big Japanese manufacturers, and it was only a matter of time before the more exclusive styling houses brought in styling by computer. The design concept, however, remains a cerebral thing, and styling--by-robot and computer is always going to be subservient to the artistic vision of a top stylist, whose intuitions and creative genius can never be replaced. Modern technology allows this inspiration to soar to even greater heights.

**CLOCKWISE FROM
ABOVE** Drawings for 1941
Chrysler Thunderbolts.
Buehrig's pastiche Duesenburg
towncar of 1949. Derham-
bodied Chrysler convertible. A
study for the Facel vega.
Perfecting details on 1949 full-
size Clay Ford. GM design
department preparing full-size
drawings.

RIGHT GM design studio,
1957. Dodge studio clay
modeller Tom Pelky makes
revisions to 1956 model.

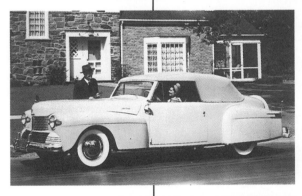

CLOCKWISE FROM ABOVE Chrysler K310. It came from somewhere else: Chrysler's answer to the flying saucer. Testing aerodynamics at Chrysler. The boys who built the turbine. 1941 Lincoln Continental. 1959 Ford Thunderbird.

THE STYLISTS

HARLEY EARL

RIGHT Harley Earl with a
1927 Cadillac LaSalle.
BELOW RIGHT Special
prototype moulds for 1937
Buick. **BELOW
LEFT** Brochure illustration
for the 1937 Buick 8.

Alfred P. Sloan, founder and Chairman of General Motors, wrote a memo to his staff in the early 1920s, drawing attention to the fact that automobile engineering had stabilized to such an extent that the only way of encouraging the public to buy a new car would be by changing the shape of it. Thus began the concept of built-in obsolescence; making car styling fashionable was the name of the game.

Accordingly, Sloan set up an 'Art and Color Section' at GM, making good use of the Du Pont chemical company's newly introduced Duco synthetic paint range. Harley Earl, known in the trade as 'Misterl', was hired by his friend, Cadillac President Larry Fisher in 1926, and he set about transferring his own influences (Hollywood and the Santa Monica race track in particular), onto the cars GM wanted to sell. The first vehicle he styled was the 1927 Cadillac La-Salle, lower and longer than any other US car, and owing much to Hispano Suiza. At the time, Cadillac had a lower profile than Packard, Pierce and Peerless, and the new look did much to lift its reputation. There needed to be a stylistic identity not only for each different model, but also for each Division within GM. The technique was to turn the public's fantasies and obsessions into reality in the shape of automobiles. Thus GM could be sure of selling 5 million cars a year.

Earl was born in Hollywood in 1893, into a family business which built horse-drawn carriages, before progressing in 1908 to the Earl Motor Works. Invalided out of Stanford University, Harley Earl began building self-designed car bodies, and sold them to movie stars like Jack Pickford and Fatty Arbuckle.

Earl's neighbour in Hollywood had been none other than Cecil B. DeMille, he of the overstated film epic: 'Misterl' brought this theatrical Hollywood background to bear, borrowing the modelling clay often used on film sets. With this

TOP Prototype of the 1937
Buick 8. **LEFT** Harley Earl
poses in the 1938 'Y' Job.
BELOW Earl's fascination
with jet fighters is clear in the
1951 Buick LeSabre.

23

medium, the stylist could develop all kinds of flowing lines and curves and see them develop in three dimensions. There was an immediate advantage over the two-dimensional concepts sired on the drawing board, and although the initial ideas were worked out on paper, the major part of the styling excercise was done in clay. It was not long before this method of styling was used throughout the whole car industry.

After a while, his methods of getting things done around the studio became a shade notorious. Sartorial and charismatic in the extreme, Earl was an intimidating 6' 4" tall, and had a reputation for being somewhat authoritarian. When his assistants had produced a finished clay, it was submitted for management approval. If all was in order, a fibreglass model was made, prior to a full-size prototype.

Another of Harley Earl's significant contributions to the US motor industry was his introduction in the 1940s of the 'Motorama'. Now Earl was also a fan of Al Jolson, and credited *The Jazz Singer* with teaching him how to judge an audience. He resolved to give the public something to whet its appetite for fabulous cars. There was no US motor show as such at this time and the 'Motorama' took the form of a lavish travelling circus, complete with live music and Busby Berkeley routines, which went from

town to town displaying Harley's and GM's dream-cars. Not only did they serve a public relations function and afford a way of entertaining motor industry personnel, but the public's perception and opinion of the latest styling ideas could be gauged. These creations included the astonishing Buick LeSabre of 1951, which exemplified Earl's other obsession: jet fighters. The LeSabre was almost a caricature of the contemporary Sabre F-86 fighter plane, featuring a wrap-round windscreen and a variety of spectacular air ducts which did nothing to assist the cooling of its humble Buick running gear. The LeSabre was his personal transport for a while, and even General Eisenhower borrowed it to impress the Nato top brass in Paris. Earl's love

LEFT Earl with his 1959 project, the blister-cockpit Firebird III turbine car.
BELOW The 1954 Firebird I, a jet on wheels.
BOTTOM Earl's finest, the 1953 Chevrolet Corvette.

affair with aircraft took off at Selfridge Airforce Base in 1943, when he first saw Kelly Johnson's Lockheed Lightning P-38. This plane was unusual in that it had a twin fuselage and twin tailfins and rudders, and it was from Earl's adaptation of these key features in his 1948 Cadillac Sedanet that the whole US auto industry derived its fins during the following decade; likewise, the wraparound windscreen, which was originally made, not without some difficulty, by Libbey-Owens-Ford. There was no mathematical formula for defining the angles Earl prescribed. After a while, everyone had to have a wraparound windshield, and the more expensive the car, the steeper the angle of rake and curvature on it.

Named after a naval vessel, the Corvette was the first of Earl's Motorama dream cars to make it to production. In terms of 'dream car', it was only dream-like in its body-styling. Externally it looked the part, combining jet-fighter imagery with the flashiest of European sports car styling. Underneath though, it harboured all the creature discomforts of the MG and Jaguar imports,

TOP 1955 Chevrolet BelAir is cleaner looking than '53 Cadillac Convertible, right.
BOTTOM Harley Earl and his 1938 'Y' Job, Firebirds I and II, plus 1951 Le Sabre.

MADLER 5-14-56

13692

plus the barge-like handling and tyres of the average American truck. The enthusiast could hot-up a motor if he wanted to, but the average forties and fifties US car buyer was only interested in how the thing looked, not how it went. Therefore it mattered little if the mechanicals were under-engineered and basic. It was the size of the fins and the amount of 'chronium' (an Earl-ism) which were important. The 1950s were the halcyon years of US car design, and many of Earl's creations epitomize them. Cars like the 1953 Cadillac Eldorado, the '53 Buick Skylark, the '55 Chevrolet Bel Air, and of course, the '58 and '59 Cadillac Fleetwoods. By this time, he was on a basic retainer of $130,000, and retired in 1959 to set up his own styling house. He died a decade later in 1969.

Harley Earl's influence was all pervasive. No area of the car's body was safe from the Earl touch. As roof lines got progressively lower, there was this vast area of plain bare metal for all to see. What could be done with it? After all, its primary function was to keep the rain out. Of course! Misterl put grooves in it!

■ When 'chronium' was available in mountains, Earl was there with a shovel, and the '58 and '59 Cadillac Eldorados got their share. Sills and valances of Biaritz model, below, are covered with it.

BILL MITCHELL

In his 17 years as vice president of design at General Motors, Bill Mitchell was responsible for the shape of some 72 million automobiles, which is getting on for 50 per cent of the whole US market. These included the '63 Riviera, the '67 Eldorado, the '70 Camaro, the '75 Seville, and the '77 Impala. However, what made his name was the star of 1963, the Chevrolet Corvette.

His father had been a Pennsylvania Buick dealer, so cars were always around. After graduating, he went to art school at night in New York, and worked in Colliers advertising agency by day. He raced home-built cars with the ad agency chairman's sons, amd one day his sketches of these activities were brought to the attention of Harley Earl. In 1935, aged 20, he joined Misterl's Art and Color Section at GM and was head of design for Cadillac by 1938, where his first success was the Cadillac 60 Special. The layout of the department was open plan, with stylists for each car-division screened off from each other. The screens didn't reach the ceiling however, and we are told that when Earl was castigating some luckless draughtsman, paper planes or lumps of clay would be dispatched over the top of the screens to add to the harassment. Such was life in the heady days of the Art and Color section!

William L. Mitchell at his drawing board as Vice President, Styling, at General Motors. **RIGHT** A model for one of his ideas in 1940, and below, the Cadillac 60 Special Sedan of 1938.

■

LEFT AND BELOW LEFT
Studies of the 1940 concept car
demonstrating all-enveloping
streamlined wings. Frontal
aspect combines Deco with
Space Age.

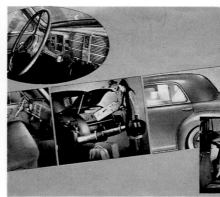

■

ABOVE AND LEFT
Extracts from the brochure for
the highly acclaimed Cadillac
60 Special.

Mitchell was always concerned to keep things looking modern, and set about making the lines of the 1941 Cadillac crisper, giving it plenty of chrome and a horizontal grille. It was soon after this that his eyes were opened to other possibilities, during a trip to California with Harley Earl. Here he saw all manner of hotted-up and cut-down cars, from Talbots and Delages to Cords and Cadillacs with special bodies. He came to realize the sterility and conservatism of Detroit.

After a spell in the navy during the war, he was made Earl's general assistant. His expertise was in motivating the design staff, who were accustomed to being cowed by Earl; instead, Mitchell joked them into a state of excitement with his swift repartee, and a creative atmosphere would be restored. There was a five year spell as head of Earl's personal design consultancy, but after that he was back as Earl's assistant once more. When Misterl retired in 1959, Mitchell got his job. His enthusiasm for the subject was all-pervasive, and he frequently referred back to his California trip of 1941, believing that car styling should be about fresh ideas and promoting an exciting driving experience. His adaptation of the Sting Ray theme in 1959, attests to this; he also sponsored the SR2 car which Dr Dick Thompson drove with splendid success during the 1960 SCCA season. Another rendition was the 1960 Mako Shark, inspired by a successful fishing trip.

The Corvette exemplified Mitchell's attitude to cars in general; his exhortation to his designers was 'Make it look like it'll do something!' It should be brash, fast, good-looking and purposeful, in short, something really special, and his 1963 Sting Ray with the split back window was just that.

After he retired from GM in 1977, he set up his own consultancy, much as Earl had done, and did a lot of work for Goodyear and Yamaha, all the while attending races and simply being an irrepressible car enthusiast.

TOP Almost a caricature of the Sting Ray, the Manta Ray looks even more menacing and purposeful. **CENTRE AND RIGHT** Two variations on the Sting Ray theme; the 1959 coupe and the SR2 racer driven in SCCA events.

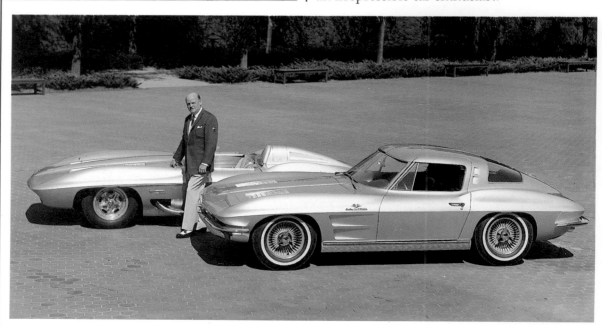

ABOVE Bill Mitchell and his Corvette SR2 sports racer, with some of GM's other sporting models. **LEFT** A selection of products which passed through GM during his time there.

ABOVE Mitchell conceived the Mako Shark after a fishing trip and adapted the Corvette accordingly. **RIGHT** Rear of the Mako Shark II coupe was as mean looking as the real thing. The similarity even extended to the gills.

■ **LEFT** Corvette SR2 on display at the Henry Ford Museum, Dearborn.
BELOW Mitchell penned the last of the classic Corvettes in 1968.

GORDON BUEHRIG

Born in 1904, Gordon Buehrig suffered as a child from the frustration of being crazy about cars, but having a father who didn't own one. On being handed down his cousin's Orient Buckboard when still a student, he set about building a speedster body for it. Having been fired as a cab driver for driving under-age at 19, he joined C.R. Wilson, a Detroit body builder. As an apprentice designer, his pay was 40 cents an hour. By 1926, after a couple of similar jobs and a spell on the road in California, Buehrig was making $200 a month at Packard as a detailer and draughtsman. At this time, General Motors was getting its exciting Art and Color Section off the ground, and the temptation to join was too much to resist, even if it meant a cut in salary. Buehrig's application was accepted.

Under Harley Earl, Buehrig's first job was designing the instrument panel for the 1929 'pregnant' Buick. But even before this model was launched, he had been taken on by Stutz, who guaranteed him a salary sufficient to pay for his new car, a 1929 Buick roadster. Buehrig stayed at Stutz just long enough to design the three boat-tailed Stutz Le Mans cars of 1929. Based on the short Black Hawk chassis, they were actually built by Weymann of Indianapolis, who were to produce many of Buehrig's designs in the future. This whirlwind career was only

just begining to take off. Stutz was getting into financial difficulties, and in the same year, at the tender age of 25, Buehrig joined Duesenburg as Chief Designer.

The state of the luxury car market was precarious to say the least after the Wall Street Crash, and Duesenburg was concerned that his wealthy customers would stray. It had been Duesenburg practice to buy in bodies from coachbuilders such as LeBaron, Murphy, Derham, Judkins, Holbrook and Willoughby, but customers could also see the same body styles on supposedly inferior makes, like Pierce-Arrow or Packard, who bought their bodies from these coachbuilders too. To maintain Duesenburg's exalted position, Buehrig's brief from Sales Manager Harold Ames was to design something really exclusive. His first offering was a close-coupled short-wheelbase coupé, built by Judkins for Schreve Archer. Next came the Beverly Sedan, which was built in some numbers, and although Buehrig was submitting a lot of designs at this time, the economy prevailed against their ever going into production. One which did was the magnificent green and yellow Derham-built Tourster, shown at the 1929 Chicago Salon, and bought by Gary Cooper. There was also the Brunn Torpedo Phaeton, built to Marc Lawrence's specific wishes in 1931,

Buehrig got the idea for the Cord whilst at the Art and Color Center, but developed the design at Duesenburg.

The inimitable 810 Cord had recessed headlamps and supercharger. The car was produced until 1936 when Erret Cord closed his automobile business.

and succeeded by several other similarly elegant designs built by Weymann and A.H. Walker.

It was lean times at Duesenburg by 1933, and Buehrig briefly rejoined Harley Earl's staff, where he got the idea for the 810 Cord. Harold Ames, however, now President of Duesenburg, had a scheme whereby cheaper Auburn parts would be used on a Duesenburg chassis to create an entry-level car. The resulting 1934 Auburn was not entirely successful, but its successor in 1935 was a rather more handsome and coherent design. Meanwhile, the earlier model was revised and re-introduced as a front-wheel-drive Cord. There had been a certain amount of juggling with the design, because the Duesenburg management wanted to use cheaper tooling and Auburn-sourced parts for the Cord, and generally compromise Buehrig's original design. Aided by Auburn President Roy Faulkner, Buehrig got the Cord passed, and with much midnight oil burned by the Auburn workforce, the Cord made it to the 1935 show.

Buehrig's other designs at this time included the Rollston 'Twenty Grand' Duesenburg Sedan of 1933, the 812 Cord, and the Auburn 852 Speedster of 1935. When E.L. Cord perceived the narrowing quality gap between his own cars and those of the mass producers, he closed down his automobile business, and Buehrig was out in the cold for some years. He designed aircraft components during the war, and worked briefly under Raymond Loewy at Studebaker. In 1949, he went to Ford as head of body development design, responsible chiefly for transforming basic saloons into convertibles and station wagons. Buehrig's 1952 Ranchwagon was the first all-metal station wagon to break with the 'woodie' concept, and in one year, it boosted Ford's estate car sales from 7000 to 140,000 units. After initial involvement with the Ford Falcon, Buehrig worked from 1959 until his retirement in 1965 as principal design engineer in the development of plastic bodies and use of plastics in automotive components.

JOHN TJAARDA

In 1951, the New York Museum of Art singled out the Lincoln Zephyr as the first streamlined car produced in the USA, and such European luminaries as Dr Ferdinand Porsche, Rasmussen of Auto Union and Dolfous of Hispano Suiza claimed it was the only car of interest to emerge from America. It's no surprise that Porsche and Rasmussen liked John Tjaarda's original rear-engined design, since it looks so much like an overgrown version of the Beetle and the post-war Auto Union-DKW. Indeed, the Zephyr is directly contemporary with the similarly-styled Czechoslovakian 4-door Tatra Type 77A, which had a rear-mounted V8.

In 1932, Ford's Lincoln business had virtually ground to a halt, and in an effort to get things moving again, Tjaarda was enticed away from General Motors' Styling Department. Here he had styled a two-door coupé with portholes and a four-part windscreen. He had already gained experience styling some crisp bodies on Duesenburg, Packard, Pierce-Arrow and Stutz chassis at Locke & Co, and on this basis, W.O. Briggs, who assembled and trimmed Lincolns, gave Tjaarda the opportunity to do what he wanted. With the approval of Edsel Ford, who formed a separate unit especially for it, the Lincoln Zephyr was developed; in some secrecy too, since the Ford board were rather sceptical of the activities of styling departments, including Edsel's own progressive efforts. Three prototypes were planned. There would be one rear-engined car, using unit-construction, torsion-bar suspension, automatic transmission and a round radiator. Secondly, there would be the same car but with the engine in front, and a mechanically similar convertible, later to become the Continental. Lincoln's chief engineer Frank Johnson developed the new 4.4-litre V12 engine, which was always dogged with unsuitable Ford components used to cut costs.

Eventually the prototype of the front-engined car and a wooden mock-up of the rear-engined car were touted around the country in a glamour show which predated Harley Earl's 'Motorama' by 18 years. Public response to the front-engined car was far more favourable than it was to the rear-engined configuration, but the wide, chair-height seats and streamlined body shape met with much acclaim. Tjaarda was able to form the opinion that there is no point in asking the public what it wants: once people see what can actually be built, they will invariably express their opinion so accurately that future trends can be based upon it.

Two test cars proved themselves sufficiently robust, which was vital, for there was much argument over the pros and cons of unit-construction at the time. To Tjaarda's delight, its contemporary, the Chrysler Airflow, turned out to be twice as heavy as the Zephyr; it had been styled by his former despairing tutor! The first year's models sold 15,449 units, more than ten times the Zephyr's predecessor.

Although prototypes for the Lincoln Zephyr were built with front and rear engine configurations, Tjaarda found the public preferred the engine at the front.

■ The Lincoln Zephyr was developed under the aegis of Edsel Ford. Tjaarda's son Tom penned the De Tomaso Pantera whilst working at Ghia.

HOWARD 'DUTCH' DARRIN

Darrin once said that he thought everything that had happened to him was significant. For instance, if the diffident Edsel Ford had taken his advice and prepared for the Model A before the end of the Model T production run, Darrin might have found himself in Detroit and been very little heard of after that. As it was, Ford lost millions in the wake of the Model T closure. Or had Dodge used a sloping windscreen on their Victory Six, as he recommended some five years ahead of the fashion, he might have found instant fortune and had little inclination to do anything else. Or had he not had a row with Harley Earl at General Motors over the alleged 'pregnant Buick' slur, things might have been different in the Depression, and so on. As it was, Dutch Darrin worked with a number of the industry's top names, like Ettore Bugatti, Andre Citroen, Louis Renault, Sir John Siddeley, Daimler-Benz, and the Panhard brothers.

He began by designing an automatic gearshift for Willys whilst an engineer at Westinghouse in 1916. After the war, he ran a flying boat service out of Miami, selling up when four of his pilots were killed. He then acquired a pair of Delages and fabricated his own bodies for them, and in so doing, met Tom Hibbard, founder of the LeBaron coachbuilding firm. The pair went to France in 1922 to set up a Minerva dealership in Paris. Much to the Belgian company's surprise they were very successful, selling cars not only to US nationals living in Paris, but to Frenchmen as well. The Hibbard and Darrin partnership continued to build bodies for other makes as well as Minerva, and they were influential in bringing about the use of steel panels, with less reliance on wood.

By 1927, they had sold off the Minerva business, and concentrated on producing bodies for a whole variety of cars, from Rolls Royces to Duesenburgs and Hispano Suizas. Darrin played polo in Lord Mountbatten's team, and as a consequence, Barkers built a Darrin-styled body for the Mountbatten Rolls. Other Hibbard and Darrin bodies destined for Rolls Royce in the States were fitted by Brewsters in New York. Their best individual customer at the time was the millionaire Argentinian sportsman Macoco, who bought the entire Hibbard and Darrin collection on their stand at the 1928 New York Show. From wondering if they would take any orders at all, suddenly they'd sold out! Hibbard and Darrin were nothing if not entrepreneurial; they even sold a Deauville château and a pedigree dog to one client, and bought and sold a castle and accompanying Dukedom in Spain. Darrin also ran an American movie theatre in Paris, showing a new film every week.

Hibbard and Darrin did some consultancy and detail work for Stutz, and Moon Automobiles, but then relations between Darrin and General Motors soured, the Depression began to bite, and Hibbard left Paris to work, surprisingly, for Harley Earl at GM. Dutch Darrin carried on his previous associations with people like Siddeley, and joined forces with wealthy Parisian banker Fernandez, who provided Darrin with a factory in which to produce his designs.

Darrin was given a commission by Studebaker, and more or less on the strength of the experience of a Hollywood party thrown by Darryl Zanuck, he moved back to America in 1937. He was soon turning out exclusive car bodies, selling them from a showroom on Sunset Boulevard (where people thought he was French) to celebrities like Dick Powell, who had a Packard 120, and Countess Dorothy di Frasso, who had a Rolls Royce. His flying experience allowed him to operate a flight training school in Nevada during the war. Darrin's next adven-

Howard 'Dutch' Darrin's most notable creation was the fibreglass Kaiser Darrin. Design features included sliding doors and sidelights which echoed the distinctive grille shape.

ture was his involvement with the Kaiser-Frazer, a firm set up by Joseph Frazer, ex-Willys and Graham-Paige, and ship builder Henry Kaiser. The stylists at the Willow Run factory were sufficiently jealous of Darrin, only a consultant, to try to block his design for the 1951 Kaiser, but Darrin out-played them, and his design was selected. Then came what must be rated Darrin's masterpeice: the stunning fibreglass-bodied Kaiser Darrin 161, manufactured during 1954. Only 435 Units were made, and Darrin acquired fifty of these for himself, fitted them with Cadillac engines and sold them from his showroom until 1958. Hollywood film star Lana Turner was one of his customers.

And so he continued, styling a Packard here, a Panhard there; one of his most extraordinary creations was a prototype for a new DKW convertible in 1956. This was known as the 'Flintridge' Darrin DKW roadster, the rear end of which featured a series of irregular curves meandering from one side of the car to the other. There were a handful of still-born projects like the one for Israeli Kaiser dealer Illian, and the scheme for special-bodied rental cars by National Car Leasing. By 1962 he was building re-styled bodies for the Rolls Royce Silver Shadow, and wondering where he could get his hands on one of the Cadillac-engined Kaiser Darrins. He created his own classic.

■
Darrin always hankered after one of his own cars, the Cadillac-powered Kaiser Darrin.

FRANK SPRING

During World War II, Hudson was engaged in building air frames for aircraft like the B29 Superfortress and other military contracts, and Frank Spring's 1941 design for the semi-unit-construction 'step-down' Hudson Commodore had to wait until 1948 to go into production. Eventually, more than 142,000 units of the 'step-down' Hudsons were built; it was so-called because the rear floor-pan was below the chassis, and there were chassis outriggers either side of the back wheels, which allowed the body line to be much lower toward the rear of the car than was normal.

Spring had studied engineering in France, and had been at the Murphy Body company in the 1920s before joining Hudson as Design Director in 1931. A real enthusiast who owned a Type 30 and a Type 35 Bugatti, Spring lived at an intense pace, flying himself around in his autogyro or riding his BMW motor bike. His office walls in the unprepossessing Hudson Tech Center building were emblazoned with pictures of 1930s female movie stars. However, Spring contrasted with his working environment, a man of aristocratic bearing, who wore tweed suits, and was evidently able to generate inter-

Frank Spring's Hudson Commodore came out in 1948 but the designs were done in 1941.

est in the projects to hand, which was difficult in a company with somewhat limited vision. Certainly he commanded the respect of his team, which included Art Kibiger, Arthur Michel, Bill Kirby, Arnold Yonkers and Robert Andrews. The team designed both exteriors and interiors, and there was no division of responsibility. All materials were ready to hand, which was a considerable convenience.

At the time of the Commodore design Hudson used plaster for the models, despite the fact that most stylists used clay. The disadvantage of plaster was that once the model was made, it was very difficult to go back to it to incorporate detail changes; more often than not a new model had to be made. It was at model stage that one of the Hudson's trouble spots was incorporated: Spring's design had a high frontal grill, which would have facilitated better radiator cooling, but the management vetoed it in favour of a low opening. One area where Spring did manage to be innovative however, was in the variety of colours he used, pioneering the use of metallics.

There was much spying on rival firms to see what new models were going to look like, and the target for Spring's design, as far as the Hudson management was concerned, was the 1942 Buick, and this is why the Hudson ended up with such a narrow back window: only 11 inches high, the same as the Buick's. Anyway, the styling team was pretty much out on its own at Hudson, and there was much disgust when the paranoid management had the 'step-down's' plaster models smashed up before anyone could record what they looked like.

TOP Spring's design team were responsible for interiors as well, as seen in this Hudson Commodore Brougham of 1949. Popular in stock car racing, the 1952 Hornet shows the low back end, made possible by the 'step-down' chassis outriggers.

RAYMOND LOEWY

Think of Loewy and you think of controversial styling, and the '47 Studebaker. Loewy was at Studebaker for 15 years from 1940 to 1955, and New York's Museum of Modern Art called his design for the '47 Studebaker one of the ten most significant design concepts in the history of the automobile. Loewy followed the '47 Champion, Commander and Land Cruiser models with a series of logical facelifts, and then in 1953 came his second *tour de force*, the Studebaker Starliner coupé. This was way ahead of anything else around. Times, however, were not good for Studebaker, which was beset by strikes and stuck with an unmodernized plant. When Haold Vance was succeeded at the helm by James Nance, Loewy's association with Studebaker came to an end.

Loewy had been a design consultant to Austin and Rootes Group in the immediate post-war period, and his company boasted over 140 clients in the US and 40 in Europe, in areas as diverse as product packaging and farm machinery. Some of his one-off designs for individual clients included the lettuce-green 'porthole' Lincoln Continental Derham of 1941, which featured gold-plated bumpers. This little ex-

travagance was actually cheaper than hard chrome-plating. He produced quite an extraordinary design for an XK-140 in 1955, which was built by Boano; influenced by a concern for efficient aerodynamics, it was spoiled by over-decoration. In 1957, Pichon et Parat of Sens, France, built a coupé on a BMW 507 chassis to Loewy's design, which sported rectangular headlights somewhat at odds with its sculpted curves. The 1959 Cadillac surely stands on its own without revision, but Loewy re-worked its front and rear to such a vast extent that it looked quite bizarre. For a start, where was the radiator grille? Keeping cool must have been difficult.

Raymond Loewy worked with Parat et Pichon again in 1960 on a Lancia coupé, which, like the Jaguar, had contours designed to promote the optimum air-flow over the car. Even the shape and placement of auxiliary lamps was intended to facilitate brake cooling, and Loewy was always ready to learn and adapt from ideas tried in motor racing. He even had an aerofoil on the back of the Lancia's roof. None of his designs was ever intended as a competition car though, a decision reflected in the opulence and luxurious treatment of the interiors.

Raymond Loewy's pièce de resistance was the 1948 Studebaker Commander.

1948 Studebaker Commander

Variations on a theme: the 1941 Lincoln Continental in convertible and saloon form. The Derham-bodied saloon has sumptuous interior.

ALEX TREMULIS

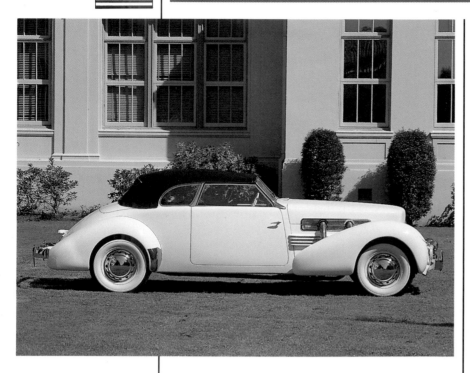

define the front wing from the rear one: the whole side of the car extended in one long, gentle sweep from front to back. The Thunderbolt had an automatic hardtop and fully enclosed wheels. For some reason, Chrysler credited the design of the car to LeBaron where it was built.

Tremulis sent off some of his aircraft designs to the US Army Air Corps as the war loomed, and was promptly posted to Wright Field air base. Here he was highly thought of as a visualizer of aircraft design, and one of his ideas was the Dyna-Soar, a vertical take-off aircraft which some see as the forerunner of the Space Shuttle.

After the war, the amiable and professorial Tremulis went to work for Preston Tucker as his stylist, having made the original cycle-wing mudguard Tucker a more viable concept. The finished result is all Tremulis, and one of his major contributions to the Tucker car was attention to internal dimensions and creature comforts. This optimistic project foundered in 1948, and for a while he was involved at Kaiser-Frazer on their 1951 model range. Like any stylist, some of his designs were speculative, and for Kaiser, he created a flashy futuristic sports car based on the Henry J. chassis. This image recurs in extended form in 1955 as a study for the Ford Mexico, a two-seater with a remarkably low drag coefficient of 0.21. From 1952 to 1963 he headed the Advanced Design team at Ford, Dearborn, penning many strange and fantastic cars and aircraft. Since then he has designed a series of record-breaking vehicles, including a motorcycle, a motor home and an Olympic bobsleigh! Alex Tremulis features in the recent film 'Tucker'; his part is played by a young actor, Eleas Koteas, who is actually a little younger than Tremulis was when he joined Tucker.

■

The Cord 812 was Tremulis' first design, completed when he was just 22. He went on to create the pontoon bodyshape of the Thunderbolt for Briggs, but Chrysler had the car built on a New Yorker chassis at Le Baron.

Hired as a stylist at Duesenburg in 1933 at the age of 19, Alex Tremulis took Buehrig's place in 1936, and his first task was to design the Cord 812. With its gleaming flexible-metal exhaust pipes exposed between bonnet and wings, the supercharged Cords were a surprise even to E. L. Cord, who thought the pipes were fakes. Auburn, Cord and Duesenburg were out of business the following year, and Tremulis moved to Oldsmobile, then Briggs, before leaving Detroit for Beverly Hills to work on coachbuilding cars for the stars. After a brief spell with Bantam's little cars, he returned to Briggs and created the Chrysler Thunderbolt, with its original pontoon body styling. This was probably the first car where no attempt was made to

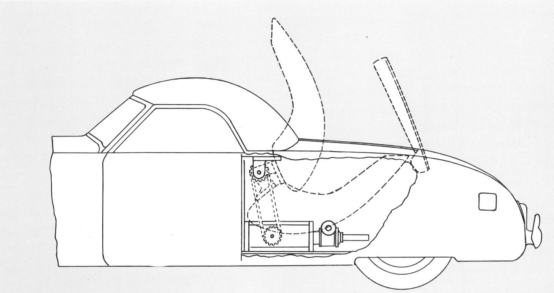

Most interesting feature of the Thunderbolt was the all-steel convertible top which folded into the boot of the car. Only six cars were built, and Tremulis went on to design many strange and fantastic cars.

VIRGIL EXNER

Virgil Exner's career took off in 1934 when he was appointed chief designer for General Motors' Pontiac division. He oversaw the 1941 and 1947 Studebaker projects, and formed the Chrysler Corporation Advanced Styling Group in 1949. From there he was elected vice president of styling in 1957, having handled the design of the 1955 Chrysler Flite Sweep and 1957 Chrysler 300C. He then set up his own firm with his son Virgil Jnr. in 1961. In 1964 he produced a series of drawings for a projected Duesenburg revival.

TOP Exner of Chrysler, Mitchell of GM, and Walker of Ford pictured in 1960.
ABOVE Rear view of Chrysler's 1955 Flite Sweep I.
RIGHT Front and rear-threequarter views of Flite Sweep II.

CLOCKWISE FROM LEFT The Chrysler 300C in production form looked somewhat cleaner than fullsize clay. Detailing of headlights was unusual. Chrysler K310 of 1952 was one product of the Advanced Styling Group.

ELWOOD P. ENGEL

Chief stylist at Ford-Lincoln during the 1950s, Engel was responsible for the 1959 Ford Thunderbird and the 1962 Ford Allegro prototype. Moving to Chrysler, Engel styled the 1963 gas turbine car, which was thought at the time to be the new direction for the motor industry. Fifty of these cars were built, and used in the main for consumer research. All but ten were scrapped. Engel also oversaw the designing of the Plymouth Barracuda fastback of 1965.

Elwood P. Engel designed the 1959 Thunderbird whilst Chief Stylist at Ford, producing the Turbine car at Chrysler in 1963.

■

**BELOW AND
OPPOSITE** Whilst at
Chrysler in the early '60s, Engel
oversaw the introduction of the
pillarless Imperial LeBaron.
Ford Allegro of 1963 has
Mustang likeness.

DAVID NORTH

David North was assistant chief designer at General Motors when he worked on the new 6.9-litre Oldsmobile Toronado, launched in 1966. This car bore many engineering and styling innovations, including the somewhat questionable reputation of being the world's largest front-wheel-drive car. Because of the fresh approach to mechanical layout, there were few constraints on styling. Apart from the Eldorado of 1967, the Toronado remained the only US representative of front-wheel-drive for almost a decade, until oil crises and preoccupations with emissions and economy brought about an industry-wide sizing-down of cars.

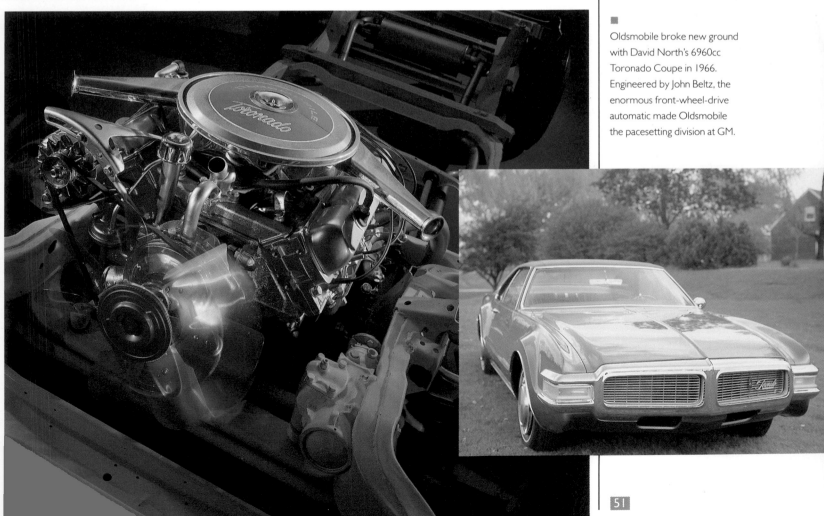

Oldsmobile broke new ground with David North's 6960cc Toronado Coupe in 1966. Engineered by John Beltz, the enormous front-wheel-drive automatic made Oldsmobile the pacesetting division at GM.

HENRI CHAPRON

Chapron coachwork on Delage D8 of 1930, the grand tourer of the day.

Born in 1886, Henri Chapron spent 60 years at the top of the French coachbuilding business, setting up his works in 1919. His speciality was rebuilding war-damaged vehicles, of which there were plenty, and in particular, Model T Fords abandoned by US troops in Paris. He fitted them with torpedo-shaped bodies, and transformed their interiors into something rather more comfortable than standard.

After the French recession of 1921, a new generation of manufacturers emerged, such as Delage, Ballot, Chenard, and Panhard, and Chapron started to make his name as a stylist and builder of certain 2-and 3-litre models. Some of his best work is to be seen in the elegant sports-bodied Delage D6 and D8. By 1927, he employed some 350 workers at his Levallois factory, and Chapron had a sizeable stand at the 1930 London Motor Show. Despite a downturn in the 1931 recession, things picked up

Chapron exhibited his version
of the Delage D8 at the London
Motor show in 1930.

when he was given a contract to build nearly all
the bodies for the Delahaye 135, which he did
until 1954, when the marque disappeared.

During the mid 1950s, Delage, Talbot and
Salmson went out of business, by which time
fewer and fewer of Chapron's suppliers were
manufacturing separate chassis-body units.
Neverthless, Chapron was not without success
during these changing times. He was responsible
for the Talbot Lago Record, the four-door
Salmson 2300, the Hotchkiss cabriolets and the
Gregoire.

After several tentative ventures into the world
of mass production, Chapron found work
modifying the Citroen DS. These he turned into
coupés and cabriolets, transforming the Citroen-
Maserati SM in the same way. During his final
years, he acted as a consultant for collectors of
vintage and classic cars.

In his later years, Chapron
transformed the Citroen DS
into the 'decapotable'
convertible.

FIGONI ET FALASCHI

Figoni et Cie. is now a Lancia dealer in Boulogne sur Seine, a suburb of Paris near Longchamp, carrying out vintage car restorations, but in the 1930s, Giuseppe Figoni and Ovidio Falaschi were literally couturiers of automobile coachwork. They would 'dress' a chassis several times before arriving at the particular line they wanted for a chassis and its coachwork. Their philosophy was that the coachwork should express the character of the chassis, and those they tended to work on were generally sporty, perhaps due to the proximity to the Longchamp racecourse. Models included Delage, Bugatti, Ballot, and Duesenberg, but it was with the Delage D8 that the firm became most associated. One of Figoni's first commissions was the body for one of the works 8C 2300 Alfa Romeos, with which Sommer and Chinetti, and then Sommer and Nuvolari, won the Le Mans 24 Hour race, in 1932 and 1933. Figoni also produced cabriolet bodies for the same Alfa chassis, but the Le Mans success led to the firm being regarded as France's leading designer and builder of competition bodies.

There is a story that Ettore Bugatti recommended Charles Weiffenbach of Delahaye to revitalize his old and tired chassis by having some interesting competition bodies made for them. This would also have the desired effect of getting some national competition for Bugatti, who was slightly out on a limb in racing. The result was that Delahaye went to Figoni for its new bodies.

The styles they specialized in were generally cabriolets, coupés, or imitation hard-top cabrio-

lets. Figoni et Falaschi were always preoccupied with the design of the car's wings, and particularly the front ones. They often went in for a type of cycle-wing mudguard, although to describe it thus is to understate its proportions completely. They were fairly substantial affairs, intended to adorn the wheel and enhance its function. Sometimes these bulbous teardrop shapes were known as 'carénées' ('streamliners'), or 'les ailes Figoni', and Figoni himself called them 'enveloppantes'. The wheel-hub was frequently fared-in, which did nothing to assist brake cooling or tyre-changing; part of the bulbous effect was to allow for at least a minimum of steering lock to be applied!

Giuseppe Figoni's original styling method was to sketch the body design on a block of paper, 16" by 24", and then reproduce the drawing in narrow strap-iron on the chassis itself. The ensuing lattice-work, or 'maquette', was supported by wood, and the sheet metal conformed to the lines and contours of the bands of strap-iron. When the 'enveloppantes' concept had become established, he began working with clay models, progressing to precise drawings. After these had been approved, his workers produced 'le grand plan', a full-scale wooden mock-up, and the sheet metal body was formed over this.

The Delahaye 18 CV Superluxe was launched at the Paris show in 1933. There was a 132" tourer with 3.2-litre straight six engine, and a shorter 116" sports model, tuned to give 113bhp. In order to popularize the new cars, Weiffenbach had Figoni design and build a single seater body

BELOW 1937 Delahaye 135 in Le Mans trim.

Figoni emerged from the war with the extravagant Lancia Belna of 1945.

One of the very last Delahayes,
the 152hp 235, which featured
a sober coupe body, by Figoni
standards..

for one of the Delahayes, and sent it off to Montlhery in May 1934 to crack long distance speed records.

The designs were initially held back by the height of the radiator, but with the introduction of the Type 135 in 1936, Figoni and his new partner Falaschi really went to town. The two-tone two-seater cabriolet had all four wheels enclosed by the Figoni-style 'enveloppantes' mudguards, and it was the sensation of the 1936 Paris show. It was bought on the spot by the Ali Khan for 150,000 francs. The reputations of Delahaye and Figoni were assured.

Another firm to benefit from the Figoni treatment was the old Sunbeam-Talbot-Darracq combine, which also went to Figoni in an attempt to revive their ailing fortunes. Tony Lago, a Director at Sunbeams, Wolverhampton, went to France in 1933 to 'troubleshoot' at Talbot. His first act was to get Figoni bodies on the Talbot-Darracq H78 chassis, and a new 3.8-litre six cylinder engine to go in it. The transformation was remarkable. The following year Sunbeam went to Rootes Group, and Lago retained control of Talbot, which he soon built up into one of France's great racing stables of the decade, alongside Delahaye and Bugatti.

F & F also built cabriolet and coupé bodies for the small Lancia Belna and Ardennes chassis, and in 1938, they designed and built a short run of ten cabriolets for Chrysler. There was a scheme to fit American Graham supercharged engines to Figoni-bodied Delahayes, and the firm also produced bodies for Delage and Hotchkiss. They patented a number of innovations, most notably the retractable windscreen, disappearing top, and compound curve 'enveloppantes', all of which were seen in the Delahaye V12 at the New York show of 1939.

Figoni et Falaschi spent the war making electric fires, emerging in 1945 with extravagant designs for the Lancia Belna, and the new 4.5-litre Delahaye 175. French popular singer Charles Trenet was to be seen in New York in a 'Narwhal nosed' Delahaye in 1948. That year, a new departure for Figoni was seen in the flush-sided pontoon bodied Talbot, which had a pair of splendid curves in the sloping rear wings.

The French tax on luxury cars of the kind Figoni and Falaschi built was beginning to make itself felt, and they tried to develop a market based on convertible-bodied Simcas. They bought chassis from Simca in lots of fifty, and built roughly three cars a week. These sold rather too well for Simca's liking, for they also had their own Facel-bodied sports car, and after some 400 Figoni-bodied Arondes had been sold, Simca called a complete halt to the arrangement.

This virtually spelled the end for the company, and Falaschi returned to Tuscany to be an hotelier. Figoni carried on, building an attractive shark-nosed car in 1952, based on the Citroen 15CV, and he did a number of Delages and Bentleys. One last try with Simcas failed, and he turned his attention to selling and servicing cars. He died in 1978, aged 84, having created 700 of the finest car bodies before the war, and some 450 in the post-war period.

SAOUTCHIK

It was not uncommon for Jacques Saoutchik to receive orders from crowned heads of state. One came by telegram from the King of Siam. All it said was: NEED THREE CARS. ONE FOR TOURING, ONE FOR OPERA, ONE FOR WIFE. DO YOUR BEST, SEND BILL. In this case, it took just five months to supply the royal ruler with a Mercedes-Benz for touring, a Rolls Royce for the opera, and a Hispano Suiza town coupé for his wife. The palace accountant was supplied with a bill for $60,000.

Not everyone is as fortunate as the King of Siam, or King Saud of Saudia Arabia, who had no fewer than eight Saoutchik-bodied cars, in-

cluding two Rolls, two Talbots, two Daimlers, and two Cadillacs. But Saoutchik was more than happy to build vehicles which were not exactly compromises, which could however look quite at home in a variety of situations. He claimed to be the originator of the convertible concept in 1928, with a design which was a limousine when closed, but could be changed into an open sports car by a system of folding top and side screens, quite a long process.

Saoutchik had set up in a Parisian side street in 1906, and before World War II it was not uncommon for clients to ask for at least two bodies for each chassis, and motorists could arrive at Saoutchik's workshop and buy a new body off the shelf. In those days, a chassis cost perhaps four times the price of the body.

By 1929, when Pierce-Arrow merged with Studebaker, Jacques Saoutchik was asked to go to South Bend and design the new Studebaker. These were good times for the French carrosserie, and Saoutchik sold around 70 cars a year at roughly $25,000 each. The orders varied. Sometimes a customer would be specific about the chassis he wanted, and the budget he could afford, and sometimes Saoutchik was given carte blanche.

Saoutchik and his son Pierre, who took over the running of the business around 1950, came to be known among the rich and famous for their elegant and highly individual designs, together with first-class build quality. Such was Jacques Saoutchik's patience that he was prepared to do as many designs as necessary in order to satisfy a customer's particular requirements. His use of chrome during the early 1950s was not merely pandering to fashion; in cars like their Talbot or Cadillac creations it became a part of the styling exercise itself. If anything, Saoutchik was ahead of the game in terms of use of chrome. He used it not as mere covering for ancilliaries like bumpers, but in great sweeping bands to define the shape of the wings and wheelarches. Saoutchik creations were nothing if not flamboyant, the 1947 Delahaye Type 175 for Laud Gaul being among the most striking. This followed the Figoni example of faring in the wheel-hubs with spats. His designs for Pegaso included one in which the entire curved windscreen disappeared at the push of a button.

In 1953, a car was made on a Cadillac chassis for one of the sons of King Saud. The order specified a convertible, big enough to accommodate seven or eight people. Saoutchik realized that no ordinary chassis would do, so he used the chassis of a Cadillac ambulance! To ensure this monster's proportions looked right, the hood-line was lowered almost to the level of the air filter, and the wings came from front to

back in a straight line before falling smoothly away, without a hint of fins. There was a glass partition between chauffeur and passengers, and a power hood was fitted. Such was the car's size that the rear seat could be turned into a bed if necessary!

The upholstery of the cars received the luxury treatment too; seat squabs, backrests and armrests were done in thickly padded leather, and door panels often carried Art Deco motifs. A mark of Saoutchik's attitude to the customer was the supply of appropriate paint colour which was dispatched with each completed order, along with additional yardage of hide or upholstery, duplicate door handles and instrument knobs. This was in addition to his guarantee that the body would last as long as the chassis. As a final touch of class, the customer could specify luggage which would match the upholstery as well.

A Hispano Suiza V12 Town Coupe of the kind ordered by the King of Siam for his wife.

BRASSEUR

Facel (Forges et Ateliers de Construction d'Eure et Loire) began making Dyna and Aronde car bodies for Panhard and Simca respectively, after World War II, turning out more than 100 bodies a day by 1952. The company also designed and built coupé bodies for the Mk IV Bentley and Ford's Vedette Comete, which was produced in some volume. This car was designed by Facel's own stylist Jacques Brasseur, and the Comete came to be thought of as an affordable version of Facel's own robust coupé, the 4.5-litre FVA, launched in 1954. The Facel Vega was probably the last of the French Grand Routier cars, for it epitomized the ideals of the breed: grand and ostentatious, elegantly styled, and powered by an engine large enough to provide effortless performance for serious long-distance touring. The company stayed clear of competitions, but development driver Lance Macklin achieved over 150 mph during high speed tests.

The frame was made of substantial tubular side rails, braced with diagonal tubes. Suspension was by coil springs and wishbones at the front, with semi-elliptic springs at the rear; brakes were less than adequate 11 inch drums. Brasseur's interior for the Facel Vega was plush, although the walnut capping was simulated. The key feature was the central console, onto which anything which overflowed from the dashboard was located, from light and wiper switches, to clock and glovebox. Electric windows were available, and after 1962, Borrani spoked wheels were fitted as standard on the Facel II. Other evolutions of the Facel Vega included the 4-door Excellence of 1957, the massive HK 500 coupe of 1959, which used 6.4-litre Chrysler power, and the smaller Facellia III, which used the Volvo 1800cc engine. Sadly, Facel went broke in 1965. All the cars they produced are worth collecting.

■ The massive Facel Vega HK 500 coupe of 1958 was powered by a 6.4-litre Chrysler engine.

Jacques Brasseur designed Facel's robust 4.5-litre FVA coupe in 1954. With the later HK 500 Facel Vegas represent the last in the line of the great French Grand Routier machines.

JEAN BUGATTI

Jean Bugatti's father, Ettore, built his own car at the age of 20, in 1901, in Milan. Before his 21st birthday, he struck a deal with Baron de Dietrich to work as a consulting engineer to produce a De Dietrich Bugatti. In 1909 came his Pur Sang racing car, and in 1910, the Molsheim factory was set up in a disused dyeworks. By 1922, when racing recommenced after World War I, Bugatti's straight-eight cars were entered in the Grands Prix, and were fitted with strange cigar-tube bodies. The following year saw an experiment with streamlined 'tank' bodies. But the best known of Ettore's designs, the Type 35 Bugattis,

ABOVE Bugatti Type 44 Fiacre. **BELOW** Type 55 exemplifies the thoroughbred post-vintage sports machine. **BOTTOM** Type 57 Sportwagon of 1935.

made their debut at Lyons in 1924, and proceeded to carry all before them during the mid-to-late 1920s; along with the P2 Alfa Romeo, they were the most successful racing cars of the decade. The styling of the Type 35 was simple and delicate; the Bugatti trademark was the inverted horse-shoe radiator, and the tail was elegantly pointed. All the exposed steering arms and suspension were small and light, and the driver sat low down behind a louvred bonnet.

From 1929, the fortunes of the Bugatti marque rested on the shoulders of Jean, (or Gianoberto), the elder son of le Patron, Ettore. Jean had been educated at Milan and Paris, and was an accomplished engineer. It was he who suggested using twin overhead camshafts for the engines which powered the Type 51, based on Miller practice. He was also a first-class stylist, responsible for the Fiacre Type 40 of 1928 for his sister Lidia. He also originated a number of streamlined designs, being the chief influence in the shaping of the Types 50, 55, 57 and 57S, and 64. A stylistic progression to the Atlantic Coupé is evident, from the streamlined Type 50 by way of the 57S to the 64. Perhaps more than any other car, the Type 55 epitomizes the thoroughbred post-vintage sports machine. And there is no mistaking the family pedigree.

Jean also designed the splendid Royale Roadster and was responsible for Bugatti's racing programme during the 1930s. This period included the 'tank-bodied' Type 59s, which won Le Mans in 1936 and 1939. Although he had been expressly forbidden to take part in competitions by his father, Jean achieved some personal successes in hill climb events. Sadly, he met a premature end in August 1939, while testing a Type 59 sports racer; his friends had blocked off a stretch of road for the testing, but they had reckoned without a drunken farmer on his bicycle. Jean Bugatti swerved into a tree and died instantly. An extraordinary talent was extinguished.

Ettore meanwhile had retired to Paris in 1936 to carry on designing rail cars and steam engines. The largest vehicles built in the Bugatti works were the rail cars made for French Railways, and powered by either two or four Royale engines. They had a fine reputation for speed and comfort, and held several speed records; the control cabin in the streamlined nose section pre-dates the French TGV by fifty years.

When World War II came, the Molsheim factory was occupied by Germans, and Ettore continued his work in Paris. His Italian nationality conspired against him, however, when he attempted to repossess the factory after the war, and despite winning a lengthy legal action in 1947, his spirit and health had failed and he died the same year, aged 66.

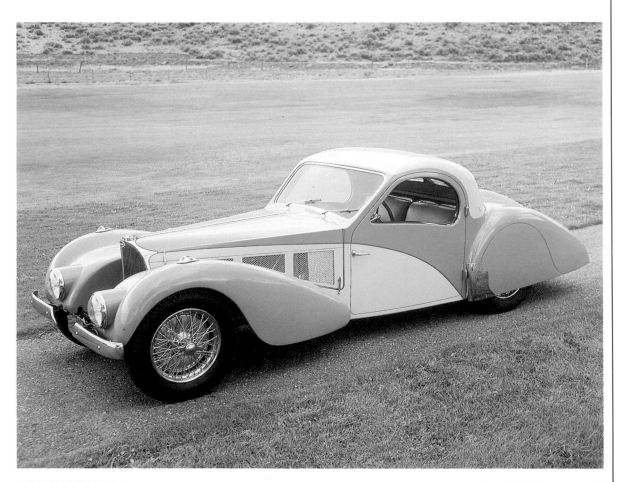

Most elegant of Jean Bugatti's creations, the 1937 Type 57 SC Atlantic Coupe. The convertible lacks the fared-in headlights.

GIOVANNI FARINA

Giovanni Farina studied the art of the carrozzeria under Peidmontese bodybuilder Marcello Alessio, setting up on his own with the Stabilimenti Farina in Turin in 1906. He gradually built up a list of well-to-do clients, providing bodies for such makes as De Dion, Itala, Sepollet, Peugeot, Diatto, and later on, Lancia, for whom he was later to do much of his best work.

He was joined by his younger brother Pinin when he left school, and the Farina concern produced military vehicles during World War I. By the early 1920s the firm was producing panels from hand-beaten aluminium, and a decade later, a number of pleasant, but on the whole fairly conservative body styles were produced for the 1934 Lancia Dilambda, the 1936 Lancia Asturia Cabriolet Aerodinamico, and the Cabriolet Royal for the Fiat 1100 of 1937. The 1939 Lancia Astura with its 4-seater cabriolet body, or the 1940 Fiat 2800 Cabriolet, were possibly the most striking of Farina's creations.

After World War II, Giovanni's son Nino enjoyed considerable success in Grand Prix racing , winning the World Championship in 1950 at the wheel of an Alfa Romeo 159; the coachbuilding and styling business continued with the striking Lanci Astura cabriolet, its flowing lines enhanced by wheel-covering spats front and back. In 1948, he made a coupé body for a Fiat 1100, reminiscent of brother Pinin's Cisitalia 202, and in 1949 and 1950 came two Ferrari convertibles. Stabilimenti Farina, however, always lacked the originality and flair of its junior sibling, and by 1952, had gone out of business.

■ Giovanni Farina styled some remarkable cars before going out of business in 1952. Here is the Lancia Astura Cabriolet of 1947.

GIOVANNI MICHELOTTI

Michelotti served his apprenticeship under Giovanni Farina at the Stabilimenti factory, which he joined in 1937, aged 16. Within two years, he had drawn the Lancia Astura. He was already well versed in the ways of the motor industry, since his father had spent 40 years in the machine shop at Itala, and could boast that he had machined the crankshaft of the Itala which won the 1907 Paris-Peking race.

Having served with the Alpine troops during the war, Giovanni set up on his own in 1949, and worked freelance for a number of carrozzeria, including Vignale, Bertone and Ghia, as well as being a consultant to several manufacturers. Notably, these were OSI, for whom he styled a Fiat spider, and Triumph, who went on to produce his Herald, Vitesse and TR4; for BMW he designed the 1500-1800 ranges, and the Contessa for Hino. He also provided the design for the Alpine A106 of 1955, fore-runner of the delectable Alpine-Renault A110. One of his styling quirks was the 'pagoda' roof-line, where the sides are higher than the centre. This was first seen on the 1960 OSCA, and subsequently taken up by Mercedes for the 230SL coupé. Another of his ideas was the aerodynamic

Michelotti did several of Triumph's family cars like the Herald, as well as sporting machines like the Renault Alpine rally car.

spoiler set in front of the windscreen on several sports racing cars of the early 1950s. He had a spell of being keen on warning lights, and was possibly the first to place trafficators on the sides of his designs. There was even a notion of mounting brake-lights on stalks on the roof for better visibility, and his quest for greater safety in car design led to cushioned fascias, counter-sunk knobs and switches for the interiors. He took streamlining seriously, consulting Professor Wunnibald Kamm of Munich, though the much discussed concepts of cut-off tail and fared-in headlights seem never to have actually materialized on a Michelotti design, however.

In 1959, Triumph announced the angular Herald, quite distinguished in its day, and certainly endowed with an incomparably good turning circle. This was followed by the 4-door Triumph 2000, and the best of the post-war Triumph sports cars, the well-proportioned TR4. In 1961, Michelotti produced what may well be his masterpieces: his designs for the Maserati 5000, produced for Ghia, following the success of Maserati's 'Shah of Persia' model. The nicest of these were the cars built for US racer Briggs Cunningham.

The Michelotti workshops are today at Beinasco on the edge of Turin. The last car to come out of the studios before Giovanni Michelotti died in 1980 was the Reliant Scimitar SS1. Since then, styling has been by Tateo Uchida, with Edgardo Michelotti taking his father's place as head of the company. The latest design to emerge is the bubble-roofed Pura, a dumpy-looking 1.8 Alfa turbo-powered design engineered by Giorgio Stirano of Albatech.

BMW's range of pedestrian-looking saloons from the 1960s was designed by Michelotti. One of his last creations was the Reliant Scimitar SS, a controversial design subsequently re-worked by Bill Towns.

ERCOLE CASTAGNA

The Castagna family had been carriage-builders for many years before going into automobile coachbuilding. This move was made by Ercole Castagna, who was born in 1885. He served an apprenticeship at the Mentana School of Car Mechanics and Design from 1887 to 1901. The firm's status was such that in 1906 they built the Fiat Sparviero for Queen Margherita di Savoia. On his father's death during the Great War, Ercole took over Carrozzeria Castagna and aided Italy's war effort by making ambulances and boats. His younger brother Emilio entered the business after the war, having attended Brera art school, and he was responsible for most of the company's designs until 1934. His work included the Alfa Romeo RLSS of 1926 and the Isotta Fraschinis, which were the hits of the 1928 and 1929 New York shows.

Best of these was the Commodore roadster; its side windows could be lowered within the doors. It was painted two-tone green, with silver-plated handles and snakeskin upholstery, and was such a success that ten orders were taken; possibly Castagna's first multiple sale of one body-chassis. Another design executed the same year was for the Mercedes SS chassis, and one of the five orders was for Al Jolson.

The Wall Street Crash spelled the end of the luxury car business in the US; Castagna's New York representative was ruined in hours. Almost at the same time, Emilio retired, and Ercole's three sons were brought in. Orders were scarce, and Isotta went bust, Alfa Romeo turning more to Carrozzeria Touring of Milan, but there were a few Lancia bodies to be done. Castagna bodies tended to be as much as 15 per cent heavier than those of Touring or Zagato, and their designs too elegant or elaborate for the mid-to-late 1930s, which meant that clients tended to be elderly conservative types.

Despite holding the Italian patent for the Labourdette Vutotal safety glass (in Italy it was called Vistotal), which made it possible to come up with fairly dramatic pillarless styling, things were quite gloomy for Castagna, which turned more and more to military contract work. The factory at Venegono was completely destroyed by a bomb in 1942; nevertheless the company managed to create a number of rather confused Fiat designs after the war, including a Vistotal Fiat in 1949, and a Cisitalia 202 with Cadillac tail fins. It was almost as if they tried too hard. The company was in the hands of the receiver in 1953, but Ercole Castagna carried on working until he died in 1968.

ABOVE Castagna bodied Isotta Fraschini Tipo 8A of 1933 reputed to have been owned by Rudolph Valentino.
RIGHT Alfa Romeo RL Supersport of 1925, one of the first designs of Emilio Castagna.

One of the most harmonious
body styles done for the Alfa
Romeo 1900 Coupe was by
Castagna in 1953.

MARIO BOANO

Born in Italy in 1903, Felice Mario Boano was a keen and accomplished art student and bronze caster. After graduating, he joined Giovanni Farina's Stabilimenti Farina as an assistant. Soon he was a tutor at the Turin school of automobile body building, and was elevated to become Farina's chief engineer. He stayed with the Farina brothers until 1934, when he left Pinin Farina's breakaway company to set up on his own. His speciality was his use of the wooden frames in the styling process, and Boano was quickly receiving sub-contracted work from Pinin Farina, Ghia, Castagna and Bertone.

When Giacinto Ghia died in 1943, Boano was invited by the family to take over the running of the company. Here he styled such classics as the Lancia Aurelia B20 and Alfa Romeo Giulietta Sprint; Ghia lacked the capacity to build these cars, and they went to Pininfarina and Bertone respectively. Boano's arrangement at Ghia lasted until 1952, when he was more or less ousted by the commercially-minded Luigi Segre, and he left to re-establish his own business.

Among his creations in the mid 1950s were the Chrysler-powered Corsair, built on a Nardi tubular chassis and shown at the 1956 Paris Show, a 2-litre Alfa Romeo Berlinetta with a huge, curved, split rear window, built for President Peron of Argentina, and the Indianapolis, built on a modified Lincoln chassis, and featuring large air intakes on its flanks. He also did work for Simca and Chrysler, and with his son Paolo, he set up the Fiat Styling Centre and design school.

Most of Boano's best work was done in the mid '50s; here is the Alfa Romeo 1900 Primavaera of 1956.

Mario Boano's designs for the Alfa Romeo 1900 Super Sprint of 1955 and 1956, the later car having fins and lacking the greenhouse rear window.

LEFT Alfa Romeo 1900 SuperGran Luce of 1955.

ALFREDO VIGNALE

Luigi Villoresi once somersaulted a Vignale-bodied Ferrari at over 100mph, and the car only sustained minor dents. Maybe Villoresi was lucky to get away with bruises, but his experience helped to consolidate Vignale's reputation which was founded on the strength and rigidity of his steel and aluminium bodyshells. Such was his confidence when designing car bodies that he spurned the clay model, preferring to beat out his metal panels by hand on an anvil, judging the curves by eye.

Together with his brothers Giovanni and Giuseppe, Alfredo established a metal-working business in Turin in 1946, and a special-bodied Fiat Topolino caused sufficient interest for Lancia to commission a body for the Aprilia.

Almost overnight, Vignale was ranked amongst the leading Italian coachbuilders, and orders came in quickly from Alfa Romeo, for whom they styled the 230bhp Carrozzeria Sport Speciale, and Maserati for the 1951 6AG; there were the Ferrari 212 and 340 America coupés, and the more restrained Lancia Aprilia Berlina. There was a Fiat 1500 coupé with spatted front and rear wheelarches, and a Fiat 1100 cabriolet with an Alfa-esque grille arangement and spats over the back wheels. By 1954, Vignale was producing about 1000 cars a year, including a Rolls Royce and a Fiat 8V many of them under his own name on Fiat 600D and 1300 chassis. Probably Vignale's best work was the fabulous Maserati 3500 spider from the early 1960s, of which 242 were made.

In 1969, Vignale's Grugliasco factory was taken over by Carrozeria Ghia, and three days later, Alfredo was killed in an accident in his Maserati. The firm continued to build cars for Ghia, including the De Tomaso Pantera, until 1975.

BELOW Amongst Alfredo Vignale's best work was the Maserati 3500 of 1960.

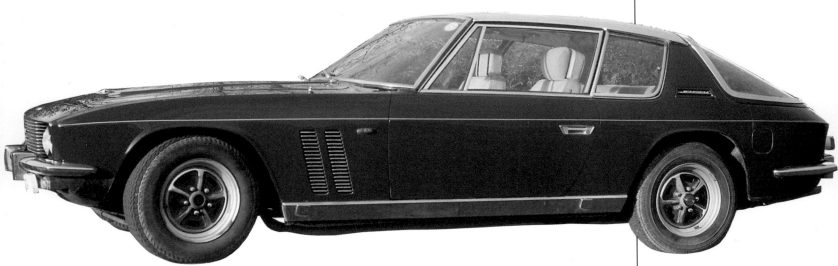

TOP Conservative '60s convertible, the Lancia Flavia Vignale.
ABOVE The Vignale-styled Jensen Interceptor of 1966 is still in limited production after a 12-year hiatus. **LEFT** The 3500 of 1962 was one of the more numerous Maseratis.

CARROZZERIA TOURING

Felice Bianchi Anderloni founded Touring of Milan in 1926. Up to this date, he had enjoyed a particularly close relationship with the founders of Isotta Fraschini in that they were all his brothers-in-law. However, family commitments restrained him from indulging in his passion for racing until after control of Isotta passed out of the family in 1922. He began racing at the ripe old age of 40, together with test driver Bindo Maserati. Having got this out of his system, he acquired Carrozzeria Falco of Milan, and, changing the name to Carrozzeria Touring, he started building his own car bodies, the anglo-appelation 'Touring' reflecting the cachet of certain English products in Italy. Anderloni began to advertise his services in the Italian specialist press; his first body was, naturally, on an Isotta Fraschini, the Tipo 8B Flying Star of 1927.

As a racing man, Anderloni was aware of the importance of lightweight construction, and built his cars on the Weymann system, for which he had the local licence. He began a long working association with Alfa Romeo in 1927, producing bodies for the supercharged Gran Sport

and Super Sport 6C 1500 and 6C 1700s. These cars were in direct competition with Zagato-bodied spiders, and often proved more effective in racing because of greater comfort and durability, although the Zagato cars would have undoubtedly been lighter. About the toughest sports car race then was the Mille Miglia, and Touring-bodied Alfas won it in 1930 (Nuvolari, 6C 1750 GS), 1932 (Borzacchini, 8C 2300), 1937 (Pintacuda, 8C 2900), and 1938 (Biondetti, 8C 308).

During the 1930s, Anderloni evolved his 'superleggera' construction system, which was basically a spaceframe chassis composed of small-diameter steel tubes supporting a sheet-aluminium skin. Body panels were attached by small clips which fastened to the tubing. Thus the body panels were completely unstressed. The Superleggera (super-light) title was used specifically alongside the company name from 1937. Although he was basically indulging his racing obsession, Anderloni didn't ignore road cars. There followed a series of gracious and refined saloon cars, rejoicing in the names of

The long bonnet conceals the straight-eight engine of the Alfa Romeo 2900B 8C of 1938.

such Latin Satanist epics as Coppa del Diavolo, Freccia di Belzebu or Soffio di Satana. These were all done on Alfa Romeo chassis, as was the bulk of Touring's work during the 1930s.

Anderloni was not of the old school, which comprised the likes of Giovanni Farina, Carlo Castagna or Ugo Zagato, who had learned their trade on the shop floor. Urbane and sophisticated, Anderloni was more alive to the fickleness of fashion, and this is why his post war designs continued to maintain their dignity and stylistic direction. Touring passed the war years as 'Turinga', a concession to Fascism, and built truck cabs. There was experimentation with light alloys, plexiglass and aircraft seats, all of which came in useful later on. The last pre-war cars were the fared-wing, flush-sided Alfa

TOP Grand Touring in fine style with the 1938 Alfa Romeo 2900B 8C. **ABOVE AND LEFT** Two of Alfa Romeo's finest cars were the 6C 2500 Sport of 1939 and the 6C 2500 Villa d'Este of 1950.

Probably Touring's best work after the War were the variations on the Alfa Romeo 1900 Sprint. Here is the version done between 1951–53, and below right, the 1900 SS of 1955.

RIGHT The Alfa Romeo Tipo 2500 6C Villa d'Este Coupe of 1950.

Romeo 6C 2500 SS roadster, which was known as the Barchetta, or 'little boat'. Both Alfa and Ferrari had used such a design in the 1940 Mille Miglia, and Berlinettas were built for BMW (Type 328) and Alfa for this shortened event. Picking up where they had left off, Touring built prototypes for Aston Martin, Bristol and Hudson. The latter, called the Italia, was decidedly American in its stylistic treatment. Further work on Alfa Romeo chassis followed, with the Villa d'Este Super Sport Coupé and Berlinetta, and when the Alfa factory was reopened, Touring was commissioned to build several thousand of the beautifully proportioned 1900 Sprints and Super Sprints. In 1948, a consortium of Lancia dealers ordered a run of elegant Aprilia Berlinettas, with the curiosity of a plexiglass occulus in the roof. Ferrari was a keen post-war customer, with the Tipo 166 and 212 Barchettas and the 24 Ore coupé all built by Touring. But Anderloni was not to see the fruits of these endeavours, for he died in 1948.

His son Carlo Felice who had been his long-time assistant, took over the business. Probably the most extravagant Touring creations were the 2- and 3-litre Alfa Romeo Disco Volantes, or Flying Saucers of 1952, which gave every impression of being made of two red saucers placed face to face; these fabulous cars with their superbly contoured wheelarches, and roofline in the case of the coupé, had an almost sharp-

LEFT Differences in frontal treatment in the Touring 1900 SS made between 1954–58. The later model got a larger rear window and overriders.
BOTTOM The original Disco Volante looked fabulous but was never a racer because of a tendency to take off.

edged waist-line. Unfortunately they lived up to their nickname, and testing at Monza proved them unraceable; however, a rather similar stylistic solution was very much more successful in the shape of Malcolm Sayer's 1954 D-Type Jaguars. The Disco Volante cars which did race in 1953 and '54 ran with different chassis, and fiercely attractive but less well-made competition bodies by Colli, whose other line was in Giardinetta estate car bodies for Alfa Romeo Giuliettas.

One of Touring's prototype offerings was the Aston Martin DB5 Coupe of 1966.

Other rather less exotic Alfa commissions included the 2000 and 2600 Spiders and the Giulia GTC cabriolet conversion on the Bertone-styled coupé. Touring's output also featured the Pegaso Type Z102, and the Maserati 3500GT series from 1959, of which 1,978 were produced, plus four of the larger 5000GTs from the same year. The list continues with the Lancia Flaminia, the Aston Martin DB4, a Lagonda Rapide, and a Lamborghini GTV in 1963. But the company's undoing was its over-ambitious move to bigger premises in Milan, and a contract from Rootes to build the Hillman Super Minx and the Sunbeam Alpine Venezia coupé. The project was a complete failure, and by 1964, Superleggera Touring was out of business.

Styling for the Aston Martin DB4 came from carrozzeria Touring of Milan. Touring also did the 4- and 6-cylinder 2000 and 2600 Spiders for Alfa Romeo.

GHIA

Carrozzeria Ghia was founded in 1915 by Piedmontese craftsman Giacinto Ghia, and his speciality between the wars was to produce bodies such as the flared-torpedo wing version of the 5-litre Lancia Kappa, for his well-heeled clients to compete with each other in Concours d'Elegance events. Other designs of note back then were for the pointed-tailed Fiat 501S and the 1929 Mille Miglia-winning Alfa Romeo 6C 1500. There was the coupé-spider body for the 2.0-litre Itala Tipo 65, and the Coppa d'Oro Fiat 508S Balilla, among others. The Torinese factory suffered bomb-damage during World War II, and Giacinto died in 1946 before the business really got going again.

The Ghia family persuaded one-time Farina stylist Felice Mario Boano to take the helm, with thrusting young Luigi Segre as commercial director. Eminent makes such as Delahaye and Talbot, ordered bodies from the factory, as well as Alfas, Lancias, and Fiats with their fashionable hinged blister fairings over the front wheels. Boano and Segre had different approaches to the work: Boano saw the car as a work of art, while Segre took the line of the industrialist, his

sights set on export markets, and it was he who prevailed. Boano left in 1950, and Segre oversaw Ghia's output until his premature death during a routine appendix operation in 1963.

There were a series of prototypes for Chrysler, including the Crown Imperials and the gas turbine cars, and work was done for Ford on the Turnpike Cruiser, for Rolls Royce, Alfa Romeo, the Predictor for Packard, and of course the well-known VW Karmann Ghia. There was also a Porsche spider, a fabulous Maserati 5000GT for Snr Innocenti, and for Renault there were the Floride, the R4 and R16 prototypes. In 1955, two 'dream cars' were produced; the aerodynamic Giovanni Savonuzzi-designed Gilda, star of the Turin Show, and the Lincoln Futura, the work of Edsel stylist Roy Brown, destined to become the basis for the George Baris 'Batmobile'. Segre had employed a number of freelance stylists, but took on Tom Tjaarda (son of John) until 1961, and Filippo Sapino to work as in-house stylists. After Segre, Ghia built a run of coupés on the Fiat 1500 chassis, under the direction of Gino Rovere. He died unexpectedly in 1964, and was replaced by Giacomo Gaspardo

FROM THE TOP Alfa Romeo 1900 Coupe, 1953. Alfa Romeo 2000 Super Sprint, 1954. Alfa Romeo 1900 Super Sprint, 1955.
ABOVE LEFT Alfa Romeo 1900 Berlina Super, 1955.
ABOVE RIGHT Alfa Romeo 1900 Super Sprint, 1955.
RIGHT Volkswagen Karmann Ghia convertible, 1968.
OPPOSITE PAGE, TOP 1987 Ford RS200. Ford Granada Altair, 1981. Ford Granada Ghia Coupe 1976.

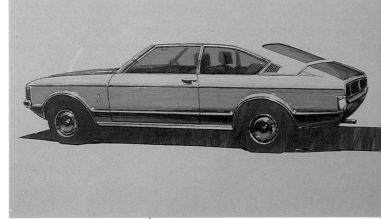

Moro. It was at this point that Alejandro De Tomaso came on the scene, commissioning Ghia to make bodies for his Fissore-styled, Ford Cortina-powered Vallelunga coupé.

De Tomaso had had a perilous career in politics and journalism in Argentina, interspersed with motor racing. By the time he went to see Ghia, he had several none-too-successful single-seater racing cars under his belt. By then a resident of Modena where his competition cars were built, he quickly developed a strong association with Ghia and acquired the company in 1967, when its Caribbean-based owner needed substantial bail funds - $650,000. Almost immediately, De Tomaso sold the controlling interest to his wife's brother-in-law, continuing to act as managing director. At the 1966 Turin show, four new designs were unveiled, the work of rising star Giorgetto Giugiaro who Moro had lured away from Bertone. They were the De Tomaso Mangusta and Pampero, the Fiat Vanessa, and best of all, the Maserati Ghibli. All was not well in the design studios under De Tomaso, however, and a number of staff left in 1967, including Moro, Sapino (temporarily),

The Ghia bodied Maserati Ghibli was unveiled at the 1966 Turin Motor Show.

and Giugiaro, all complaining of a lack of design autonomy at Ghia.

Tom Tjaarda was brought back, and penned the 'Ford-Ferrari,' the De Tomaso Pantera, most of which were built at the Vignale factory which Ghia had taken over in 1969. These could be marketed in the US at Ford's Lincoln-Mercury outlets, an arrangement made feasible when De Tomaso sold a major holding in his company to Ford USA. One of the proposals which may have attracted Messrs Ford and Iacocca was the De Tomaso Deauville, an XJ6 lookalike. By 1971, however, the autocratic De Tomaso had resigned his remaining holding in Ghia, and under Ford's supervision, there was plenty of stylistic input from the studios, coming to fruition with production models like the Granada Ghia, the Mustang II Ghia, and the V6-mid-engined AC Ghia of 1981. There were prototypes for purpose-built town cars, all-terrain vehicles like Sapino's Saguaro of 1988, and the fabulous Ford Group B rally car, the RS200. In the public eye, however, the name of the old Italian coachbuilder now represents little more than luxury trim packaging on top-of-the-range Fords.

■ **ABOVE** By now a Ford subsidiary. Ghia studios designed the Ford Mustang II.

LEFT AND BELOW Stylistic triumph of an earlier decade, the Maserati Ghibli.

ZAGATO

If the famous 'Z' emblem reminds you of a streak of lightning, Ugo Zagato would have been happy, for he introduced the world to some of the lightest, and therefore fastest, sports-racing bodies ever. He founded Carrozzeria Zagato in the Via Giorgini, Milan in 1919, and he really came to prominence with his bodies for the 6C 1750 and 8C 2300 supercharged Alfa Romeos of the late 1920s and 1930s.

The trademark of some of the Zagato designs was the so-called Panoramic look, which had the front, side and rear windows curving slightly over the car's roof-line. In 1947, his son Elio now working with him, Ugo produced a Fiat 500 and an 1100 with this styling quirk, and began to climb out of the economic mess left by the war. During the late 1940s and early 1950s many small-capacity Fiats fitted with Zagato bodies were raced, and Elio was national 750cc champion with a Deutsch-Bonnet engined car in 1952. Zagato's next step was to acquire several of the new Fiat 2-litre V8 chassis, and clad them with light, streamlined bodies. These cars dominated the 2-litre GT class from 1953 to 1955, finishing 9th overall in the 1954 Mille Miglia. The Zagato Fiat was pipped for the 1956 championship by a Zagato-bodied Maserati A6G. A more hefty looking car than the Fiat, the 1956 Maserati featured a stylized version of the company's trident mascot in its radiator aperture. Competition for the Fiats also came from Zagato-bodied Alfa Romeo 1900 Super Sprints, and 2.5-litre Lancia Aurelias.

The difference a lightweight Zagato body made to the Alfa Super Sprint was quite surprising. The standard car in 1954 was capable of just 114mph, but with the Zagato body and no engine improvements, it could top 121mph. The weight lost was 396lb. Similarly, the performance of the Aurelia was increased from 114mph to 124mph. The 2-litre Maserati A6G gained 10mph, taking it to 130mph.

Other models to appear with the Zagato treatment in the mid-'50s were the Fiat Abarth, with its extraordinary cooling ducts, and capable of 93mph in spite of its lowly 750cc engine; then there was the Alfa Giulietta Sprint Veloce 1300, known as the SVZ, and the rotund little SZ and its long-tailed sister, the SZT. These cars were regularly driven to and from circuits by a host of private entrants during the 1950s, and were more or less invincible in their class until the advent of the 1600 Porsches, Simca-Abarths and Lotus Elites in the early 1960s. Another Zagato trademark was the 'double bubble' roof, in which the roof area is divided, with either side forming a shallow dome shape. Some Zagato bodied cars like the Fiat Abarth and the Lancia Appia got the same treatment extended to the boot-lid too.

There was a somewhat ungainly Issotta Fraschini limousine with a rear-mounted V8, but excursions away from sports cars were rare. Commissions along the lines of the Le Mans prototype designed for Lotus were more up Zagato's street, and there was other work for

Zagato came to prominence with the 6C and 8C Alfa Romeos like this 1750 GranSport of 1930.

The Alfa Romeo 1900 SS and Spider of 1957–9. Early versions of the Lancia Fulvia Zagato had alloy bodies. Alfa Romeo's monster ES30 Coupe of 1989 is based on the 75 platform.

Ferrari and Maserati, which included the monster 450S Le Mans coupé of 1958. Let's not forget the big Lancia Flavia Z and Alfa 2600 Z coupés either. The Zagato family, by now with Elio's brother Gianni on board as principal stylist, was prolific in its output; by 1960, a third of the cars racing in the arduous Sicilian epic, the Targa Florio, bore the legendary Z emblem, their aluminium bodies still hand-beaten.

The Zagato method of body construction began with a normal sheet-steel floor pan, to which a rigid cage was hand-welded. This was made out of short lengths of tube, in larger diameter for stressed areas, and with pieces of shaped sheet steel forming an inside lining for the unstressed aluminium panels of the exterior. The panels were seam and tack welded onto the frame, and after painting (just two top coats sufficed in 1961!) trim and upholstery were made and fitted by skilled craftsmen.

By 1962, a new plant was on the way, just north of Milan, and Zagato went on to produce some of its finest work for Alfa Romeo in the Giulietta SZ, the TZ and the aggressively functional TZ II racing GTs. These designs reflect Zagato's basic philosophy, which calls for minimum weight and the smallest possible frontal area; ultimate aerodynamic efficiency was assumed to require too much rear overhang, so the chopped-off Kamm-tails provided a sound solution. The Giulia-derived Junior Z, and its contemporary, the Lancia Fulvia Z, came later. There were cars for Bristol, and a masterpiece for Aston Martin. The DB-4 Zagato is arguably the nicest of all Astons; and the Aston Martin V8 Zagato of the mid 1980s is one of the most purposeful-looking sports cars of all time. Continuing its tradition of working with Alfa Romeo, Zagato built the company's most recent GT car, the startling ES30.

Aston Martin's DB4 GT Zagato presented a more rounded aspect than the standard car.

Typically Zagato in its chunkiness, the DB4 GTZ is arguably the nicest Aston Martin shape. The Alfa Romeo Guilia TZ 2 of 1966 scored many class wins in long distance endurance events.

PIETRO FRUA

What is it about Piedmont which breeds talented stylists? Like so many others who started off with coachbuilding backgrounds in the auto cities of Turin and Milan, Pietro Frua was born in this Northern Italian region in 1913, a keen rugby player and skier. He was an apprentice at the Fiat mechanical training school until he was 17. Then he went to work for Giovanni Farina, working alongside such gifted colleagues as Mario Boano, taking over as general manager when Pinin Farina left the company in 1930. An ambitious young man, Frua had to try going it alone, and there followed a tough period of turning his hand to anything, from pedal cars to electric ovens; one such design was to become the Vespa motor-scooter. After the war, he set up a small factory in a bomb-damaged building, and took on 15 coachbuilders.

One of Frua's first commissions came from Maserati, to make commercially viable bodies for the 2-litre A6 GCS. Frua's solution was decidedly contemporary; there was no shilly-shallying around with references to the pre-war separately defined wings and mudguards. The sides of the car were done in one long, harmonious sweep. Its large and uncompromising grille brooked no messing; there is a real quality of timelessness about the design. Like that other

The Renault Floride, above, was an attractive convertible on the Dauphine platform. AC cars intended the 428 to be a replacement for the Cobra but the project was abandoned with only 50 cars built.

OPPOSITE The Frua studio also had a hand in the styling of the Volvo P1800S, immortalised by Roger Moore in 'The Saint.'

popular anachronism, the AC Cobra, Frua's Maserati A6GCS and A6G/54GT wouldn't look out of place at a motor show today.

In 1955, Frua came to an arrangement with Ghia, which in essence gave Ghia first refusal on Frua's drawings. The most notable vehicle to emerge from this association was the Renault Floride, a perky little coupé version of the rear-engined Dauphine. A nice stylistic solution this, but the wheels always looked too close together. Frua couldn't tolerate the Ghia overlordship for long, and he broke away amid recriminations surrounding authorship of the Floride design. His own business was called Studio Technica Pietro Frua. Designs included the Volvo P1800, partly the work of Pelle Petterson, made famous to millions by Roger Moore's use of it in the TV series *The Saint*.

Frua produced several attractive bodies for the Maserati 3500GT, and in 1963 two Maserati 5000GTs. An earlier one-off 4-door Maserati made for the Aga Khan was the basis of the Quattroporte, for which he won the Maserati contract in the same year. The panels were actually made by Vignale and the cars were assembled at the Maserati plant at Modena. This car was seen by Maserati's Omer Orsi as a means of squaring up to the Jaguar saloons, then enjoying considerable market penetration in Italy. Maserati also commissioned the linear Mistral from Frua. Records show 948 were built, of which 120 were spiders, from 1964 to 1970. Directly related stylistically to the Mistral was the AC 428, commissioned in 1965 by AC as a Cobra replacement, and undoubtedly one of the finest looking GT cars ever made. The chassis came out from England, and bodies were fitted and trimmed in Turin. Unfortunately, this process was far from co-ordinated, and by 1969 only 50 cars had been assembled. Steel strikes compounded the delays, and AC abandoned the project in 1973.

There had been a period in the mid-1960s when Frua did a lot of work for Glas of Germany, including the V8-engined Glas 2600 GT. But Pietro Frua worked better as an independent stylist, without the worries of long-term commitment and constraints of vehicle production. Among his one-offs was the Lotus Elan Frua SS of 1964, a pretty little GT car. His styling exercise for the Citroen SM bears a hint of Bora about the back end, and there was a further brilliant design for Maserati, the Kyalami 4-seater grand tourer, which remained in production until 1981. Along the way, attempts were made to re-shape the Jaguar S-Type and E-Type, as well as the Chevrolet Camaro; a VW-Porsche 914/6 was restyled in 1971 for the Porsche concessionaire in Spain, and this might have made it into limited production but for a disagreement over styling copyright. By the late 1970s, Frua had rather faded from the limelight, and he died in 1983 aged 70.

PININFARINA

His real name was in fact Giovan Battista Farina, and Pinin was a nickname meaning 'little boy' which stuck, to such an extent that by the time he was 60 years old, he had obtained legal permission to change his name to Pininfarina. From his origins in traditional personally hand-crafted coachwork, his career spanned practically the whole of the history of car styling; his name is today associated with many of the very best designs on the road.

Born in Turin in 1895, the youngest of ten children, Pinin went to work for his brother Giovanni at his Stabilimenti Farina coachworks. He left in 1930 to set up his own business on the Corso Trapani after nearly twenty years working for his brother, and was able to ride out the tough times of the depression; he visited the US regularly to keep a weather eye on the scene there, and at one point was offered a job by Henry Ford.

Pininfarina's son Sergio and son-in-law Renzo Carli controlled the business after Pinin's death in 1966 and did most of the styling in the 1950s. Now chairman, Sergio worked in the same way as Pininfarina himself had done in the 1930s: the sketch became the wooden mock-up, which became a full-size plaster model. Today, Pininfarina is employing computer-aided-design and experimenting with composite materials.

Pinin's first car was the Lancia DiLambda of 1932, followed by a Fiat Ardita, a miniature phaeton in which the back seat passengers travelled in what amounted to an enlarged 'dicky-seat.' There was a rather upright 4-door saloon on a Bentley chassis soon afterwards. Its sloping radiator and the line of its front wings hinted at sporting performance. In 1937 came his aerodynamic Lancia Aprillia, which had the fared-in wheels treatment, curved windscreen, and a tapering envelope body. The next year,

BELOW The 1956 V6-engined Lancia Aurelia B20 was a very advanced car.
BELOW RIGHT 1955 Peugeot 403 Berlina. Pininfarina has been responsible for Peugeot styling since the 1950s.

BELOW The Alfa Romeo Guilietta Spider 1300 was built between 1956–61. Similarly bodied Guilia spiders succeeded it until the Duetto 1600 came along in 1966.

CLOCKWISE FROM TOP LEFT 1966 Alfa Romeo Duetto. 1956 Guilietta Spider. 1971 Alfa Romeo 33 Cuneo, 1969 Alfa Romeo 33.2. 1989 Ferrari Mythos.

Cockpit of the 1966 Duetto, functional and attractive.
LEFT One which might have been; the 1972 Alfetta spider prototype.

Pinin Farina produced an elegant Lancia Astura cabriolet with 'torpedo' mudguards.

Many people believe that his first port-war creation was never bettered for sheer harmony and coherence. This was the Cisitalia 202 coupé, which certainly set new standards in styling for this type of car. It was built on a shoe-string and powered by an 1100cc Fiat engine in a tubular-frame chassis, which sounded more promising than its actual performance. At any rate, it was thought sufficiently highly of in the US to be included in the New York Museum of Modern Art's contemporary salon of influential car design.

Apart from consultancy work which produced ranges like the finned Peugeots, Lancias and Austin Cambridges of the early 1960s, there was nothing ordinary about Pininfarina's 'bread and butter' cars. During the 1950s, the factory built the charming Alfa Romeo Giulietta spiders, and in 1956, Pininfarina acquired one of the Colli-bodied Disco Volante racers, and proceeded to endow it with a succession of extravagant bodies. The first had a clear glass roof and clear plasic spats over the front wheelarches; long and lean, the fourth retained the clear-glass roof and was the most attractive; but it was the third

A feast of Ferraris. Enzo Ferrari went to Pininfarina for most of his body styles.
FROM THE TOP 1956 Ferrari 410 Superfast. 1957 Ferrari 410 Superamerica. 1966 Ferrari 275 GTB.

evolution that clearly provided the inspiration for Pininfarina's much adored Duetto spider of 1966, immortalized by Dustin Hoffman in 'The Graduate'.

Meanwhile, esteemed customers included Enzo Ferrari, who hardly ever went anywhere else for his bodies, and for whom the Pininfarina factory has over the years produced some of its best work, like the the 400 Superamerica. Really, any of the Sergio Scaglietti-built cars are wonderful examples: the 375 MM berlinetta Le Mans car of 1954, the 250 GT of 1956-64, or the 275 GTB4 of 1966. Perhaps your taste is for the 1968 365B/4 Daytona. The P5 prototype from 1967 and 512S of 1969 were far more streamlined than their contemporary P4 and 512 sports racing cars, and harbingers of production Ferraris a decade or so later. Similarly, the prototype spider done on the XJ-S for the 1978 Birmingham Motor Show prefigures the impending Jaguar F-Type of the 1990s.

Pininfarina is now set to be controlled by the third generation of the family, Paulo and Andrea. The latest venture from the Pininfarina

All Ferraris are race-bred, including the Berlinetta Boxer above, and the latest F40, left and below.

works at Grugliasco is the Ferrari Mythos prototype, under the direction of chief designer Lorenzo Ramaciotti, ostensibly a styling exercise based on the Testarossa drivetrain. This low-slung, clean-lined roadster is intended to combat the new wave of sports cars coming out of Japan, many of which are extremely well-styled (by groups of designers) and substantially undercut the prices of European supercars. One of these is, of course, the 1989 Ferrari 348, a design which continues the Pininfarina-Ferrari lineage; surprisingly, Pininfarina has given the Ferrari a frontal grille which proves to be an unnecessary cosmetic artifice.

Bread-and-butter cars during the past two decades were the classical Fiat 124 Spider, of which 200,000 were made, the attractive but underdeveloped Lancia Monte Carlo, and the best-known opus, the Alfa Romeo Spider. In 1988, the 100,000th Alfa Spider left the factory. Soon afterwards, Spider production was temporarily disrupted by a fire at the factory. Recent consultancy work has resulted in the widely acclaimed Peugeot 205, 405 and Alfa Romeo 164 volume production saloons, while the production lines ferry Ferrari Testarossa bodies alongside those of Peugeot 205 cabriolets and Cadillac Allantes. The new Pininfarina Extra design consultancy will consider styling any products, so long as their makers are best in their field.

ABOVE 3-wheeler Prototype 'Y' of 1961.
RIGHT AND TOP 1986 studies for Alfa Romeo based on 75 or 164 platforms.
BELOW 1957 Fiat Abarth 750 record breaker.
BELOW RIGHT Alfa Romeo 164.

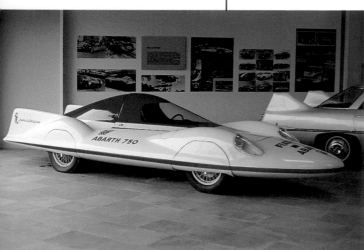

Cadillac Allantes are built at Pininfarina before shipment to the USA.
BELOW 1975 Alfa Romeo Eagle was a styling exercise on the Alfetta platform.

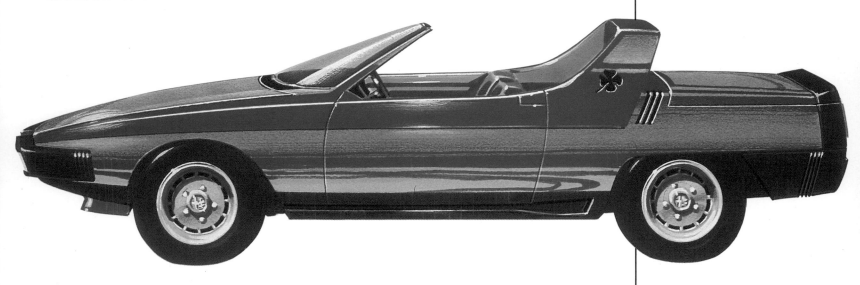

GIORGETTO GIUGIARO

Stylists are a little like musicians; they always seem to have played in someone else's band before you hear about them. Giugiaro, now an old hand with a long-established consultancy of his own, did his time with other prominent studios. First of these was Fiat Centro Stile, which he joined aged 19 after an accident involving a bucket of paint, a church altar and some unfinished frescoes of his father's. At Fiat he was part of a large team, but he learned design techniques while working on the real thing. His first break came at the Turin Show in 1958, when Nuccio Bertone saw one of his designs, and offered him Franco Scaglione's job which had just become vacant. Aged only 21, Giugiaro became chief designer of one of Italy's top carrozzerie, and his first effort was the 6-cylinder Alfa Romeo 2000/2600 Sprint coupé. This was followed by the 4-cylinder Giulia Sprint GT, announced in 1963, and which put Bertone back on the map again. In only six years at Bertone, he produced more than 20 successful designs, including the tiny Fiat 850 spider, the Fiat Dino coupé, the Iso Grifo A3C,

the Chevrolet Corvair Testudo and the stylistically related Alfa Romeo Canguro, as well as the Gordon Keeble and BMW 3200CS, which both echoed the Alfa 2600 coupé.

Giugiaro left Bertone in 1965 for Ghia. Just after he joined Ghia, the company was taken over by Alejandro De Tomaso, and although the two men didn't quite hit it off, some remarkable motor cars ensued. At the Turin Show of 1966, Ghia showed off the De Tomaso Mangusta and the Maserati Ghibli, both crisp, clean and sharply-angled designs. The following year, Giugiaro did styling exercises on an Oldsmobile Toronado (the Thor), and a 4-door Iso Rivolta, and also produced a battery-powered city car.

It was time to move on, and Giugiaro left Ghia in 1967 to set up his own Ital Design company the following year, quickly taking on a staff of around 100 as stylists, carpenters and model-makers. His partners here were Aldo Mantovani, Luciano Bosi, and Gino Boaretti, each a specialist in body structures, production methods and tool design. Giugiaro's methods of design were somewhat different from the ac-

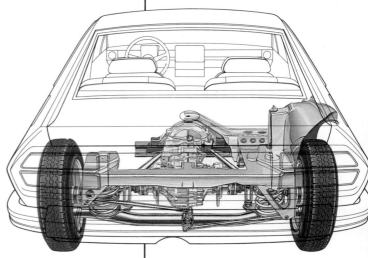

ABOVE LEFT Powered by flat-four engines, front-drive Alfasuds were swift and entertaining.

ABOVE Alfetta GTVs had clutch and gearbox mounted in transaxle arrangement.

LEFT Assembly lines at Arese produced Alfetta GTVs from 1974–88.

The Lotus Esprit in turbocharged form was good enough to challenge traditional supercars.

BELOW RIGHT AND BOTTOM The ill-fated Delorean gullwing coupe has its engine behind the rear wheels.

cepted norm. He turned an idea around extremely quickly, and did not necessarily use clay scale models, which he regarded as time-wasters, unless there was a call for a wind-tunnel test, preferring to go straight to full-size. The 1:1 model was made on a wooden frame covered with plaster. Also full-size, his perspective drawings were preceded by the plan drawings, and were air-brushed by assistants. Prototypes were usually done in steel, unless the commission was for a fibreglass body.

Consultancy work arrived swiftly at Ital Design's fashionable new Moncalieri factory, and it wasn't long before the firm had completed the entire design for the Alfasud and its production. It was the era of the wedge, and at the Turin Show in 1972, the prototype for the Lotus Esprit was introduced. The Maserati Merak and Boomerang came next, with the Alfa Romeo GTV, Volkswagen's Golf and Scirocco, all done in the same year. There was consultancy work too for Hyundai which resulted in the Pony. All Giugiaro had to go on was a photograph of the engine sent to him from Korea, but prototypes were running well inside a year. In 1975, Giugiaro styled the ill-fated Delorean, based largely on his 1970 Porsche 914 Tapiro prototype. He styled a one-off limousine too: the Maserati Medici in 1976. After the wedge-generation came the Eurobox, best example being Giugiaro's Fiat Uno, not surprisingly Italy's biggest selling car during the mid to late 1980s.

Other areas of Ital Design activity include fridges, Ducatti motorcycles, motor launches, shavers, hair dryers, furniture and bathroom fittings. By 1989, some 330 personnel were employed, and one of their offerings at the Turin Show was the Audi Quattro-based Aztec, a low-slung gull-wing, dual-cockpit sports car with lots of little scoops and louvres, which in its silver body colour looked not unlike a product of the Ital industrial design section. The ideas behind this car are logical however. Conscious of servicing constraints, Giugiaro has located all the car's servicing points behind one panel, so these functions may be attended to from outside the car. The Aztec also has an in-built jacking facility, a pneumatic wheel-brace and airline. There is considerable interest in the project in Japan, which is where not just Giugiaro, but Pininfarina too, has been selling ideas for the past two decades. These emerge unsung, not in complete cars, but in details like roof pillars or door handles. The Aztec shows there are still plenty of fresh ideas at Ital Design.

The Fiat Uno was Italy's most popular car in the late 1980s. Turbocharged versions were high powered roller skates.

NUCCIO BERTONE

Things looked bleak for Nuccio Bertone in 1952, when Fiat policy curtailed expansive freelance coachbuilding commissions. It was clearly time to stand on his own two feet, and he fitted a pair of MG chassis with coupé and cabriolet bodies, displaying them at the Turin Show in the hope that someone might pick up on them. This marked the renaissance of Carrozzeria Bertone, for they were bought by US entrepreneur and MG dealer 'Wacky' Arnolt, who visualized a market for the cars in the States. In

fact, he agreed to take a hundred of each, if the price was right. The MG chassis were transported from England, bodied in Turin, then shipped from Genoa to the USA. Arnolt finally bought some 460 bodies from Bertone, including those for the subsequent Arnolt-Bristol.

Nuccio's father was a coachbuilder in Turin from 1912, initially working on Diatto chassis, then striking up a relationship with Vicenzo Lancia in 1919. Having built Lambda bodies throughout the 1920s, he came unstuck when he turned down an offer to make all Lancia's new all-steel monocoque bodies. Lancia built them for itself, and Bertone Senior was just about out of business. Nuccio (his name is a derivative of Giuseppe), had been working for his father since the age of 12, and in 1932, aged 18 and with an economics degree in the offing, he set about selling his father's designs for the little Fiat Ballila. He was so successful that not only was the business once more secure, but the sales tour became an annual event.

After the war, Nuccio Bertone raced Fiats, OSCAs, Maseratis and Ferraris in hill climbs and events like the Targa Florio, always waiting for the body building business to recover, and all the time accumulating priceless experience about aerodynamics, cooling and balance, and how cars should be set up for competition work. Then came the Arnolt episode, and following

ABOVE Bertone's stunning Alfa Romeo 2000 Sportiva of 1954.

RIGHT The Giulia Sprint of 1962 superseded the smaller grilled Giulietta models.

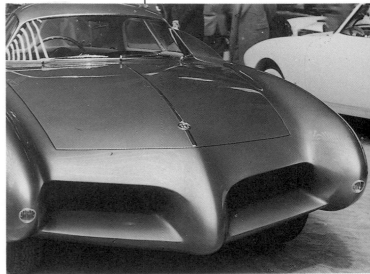

this came a design for Abarth for the 1952 Paris show. This wonderful creature had a third, centrally mounted headlight, and embryonic wings, which matured in Bertone's next project. Alfa Romeo was impressed by the Arnolt-MGs, and wanted a successor to the original Touring-styled Disco Volante. Bertone's solution was the outrageous series of BAT cars for Berlina Aerodinamica Tecnica, based on the patrician 1900 Super Sport mechanicals.

Resemblance to the vehicle used by Batman is purely coincidental, but certainly all three BAT cars displayed the most extravagant bat-like tail fins ever seen. The shark-nose of BAT 5, the first car to be made, anticipated Ferrari by

TOP The extraordinary BAT cars, numbers 5, left, and 7.
ABOVE Bertone's only Ferrari, the 308 GT4 Dino in European guise and left, to USA specification. This was the first rear-engined 4 seater Ferrari.

several years, and the retractable headlights opened and closed inward within the points of the wings. BAT 7 had the most dramatic fins of the three, but a less distinguished nose than BAT 5; in all cases the rear window sloped away to the tail in a central point. BAT 9 was a rationalization of its two predecessors. The missing numbers were accounted for by shapes which either never got off the drawing board (nos. 6 and 8) or when tested on the autostrada, failed to improve on the standard 1900 SS. Arnolt bought BAT 5, and 7 and 9 also went to the States.

In the wake of the BAT cars came a more down to earth grand tourer, the attractive Sportiva prototype, which featured a curved windscreen, slightly pointed nose and truncated rear-end. Also in 1954, Alfa Romeo was planning the 1300cc Giulietta sprint, and Bertone together with his chief designer Franco Scaglione, enhanced Alfa's own ideas to produce one of the classic GT cars, priced within reach of the regular motorist. Demand was such that Bertone needed a new plant, and by the time the Giulietta range was replaced, some 35,000 Sprints had been produced. Yet more appealing was the curvaceaous Sprint Speciale, built in far fewer numbers.

Over the years, Bertone gathered around him a number of fine assistants, and it is to Giugiaro that credit must go for the conception of the Alfa Romeo Giulia coupés in production from 1964 to 1976. The Giulia also provided the basis for the V8-engined Alfa Romeo Montreal,

■
TOP AND ABOVE The Iso Grifo of 1965–74 was powered by 7-litre Ford or Chevrolet engines.
RIGHT The Maserati Khamsin was styled at Bertone while Giugiaro was working there.

TOP Rear view of the Maserati Khamsin.

ABOVE AND LEFT The Montreal was created for the Canadian Exhibition of 1967. Cute little sports-car: Fiat's rear-engined 850 Spider was made between 1968–72. The Alfa Romeo GTA of 1965 was the competition spin-off from the popular coupe range.

ABOVE Niki Lauda discusses the 'Special Edition' Fiat X-1/9 with Nuccio Bertone.
RIGHT Neat Dallara conversion.
BELOW X-1/9 'Runabout'.
BOTTOM LEFT X-1/9s were built at the Bertone factory.
BOTTOM RIGHT 1981 Fiat X-1/9.

produced in 1967 for the Montreal Exhibition. Inevitably, the presence of master and pupil led to confusion over who designed what, rather like a Renaissance painting, where you have to decide which bits the master got his assistants to do, and which he saved for himself. No single design has been more disputed than the Lamborghini Miura. In 1965, when Ferruccio Lamborghini commissioned Bertone to design the body style for the exciting new P400 project, Giugiaro had just left Bertone, and his place was filled by the 27-year-old Marcello Gandini. Gandini had to finish the Miura designs in double-quick time for the 1966 Geneva show. It has been suggested that some drawings Giugiaro left behind for a mid-engined Alfa look strikingly like the Miura. The master, Nuccio Bertone, claims that because of Gandini's relative inexperience, he gave the designs his final touch. Who can argue with the boss?

During his 13-year stay at Bertone, Gandini was responsible for the rather sober Ferrari 308GT4, the first Ferrari to come from a stylist other than Pininfarina for 20 years. This departure from Pininfarina may have been a Fiat management decision. The 308GT4 was the first 2+2 and the first V8 in the Dino series, with a neat and uncluttered if overmodest design. Bertone himself obviously liked it: he used one as his personal transport for many years. There had been a number of restraints imposed on styling for the 308GT4, and in 1976, Bertone created the aggressively angular 308-based targa-topped Rainbow prototype to redress the balance. Gandini's Lamborghini Countach of the early 1970s showed what the studio was really capable of.

The little Fiat X-1/9 sports car was styled in 1972, again by Gandini, and due to its poor projected sales potential, it was built at Bertone's factory, rather than at the Fiat plant. It wasn't phased out until 1989, but was never seriously developed by Fiat. Bertone's project for 1989 was the Genesis, a futuristic design exercise on the Espace people-carrier theme, ending up as a cross between a camper and a Countach.

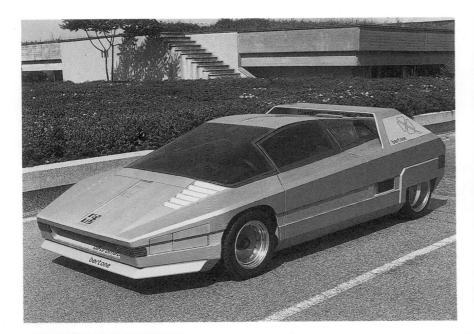

TOP LEFT 1976 Alfa Romeo 33 Navajo prototype.

TOP RIGHT The Bertone factory at Turin.

LEFT The Fiat X-1/9 was styled by Gandini when he worked at Bertone; the full potential of this mid-engined sports car was never realised.

Stylists at work on the Ramarro project in the Bertone design studios. Detailing work is done on the full-size clay.

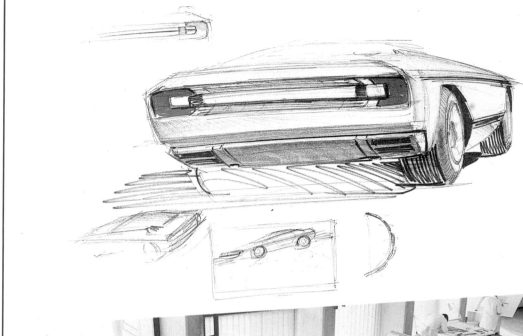

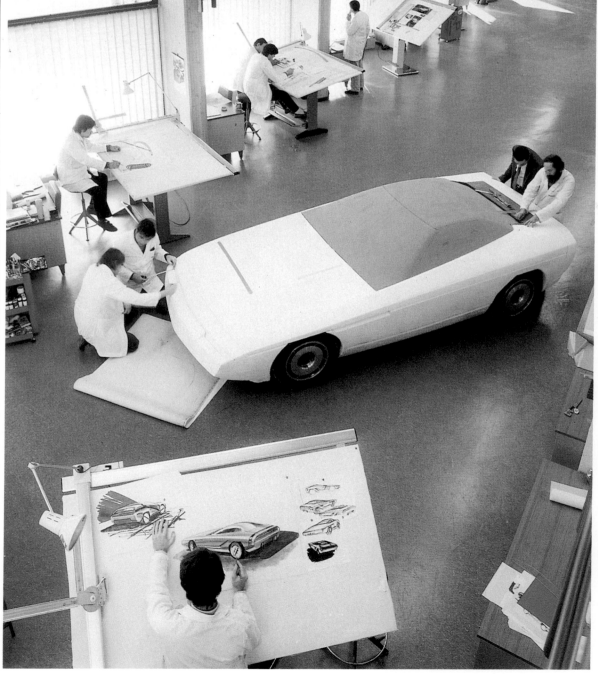

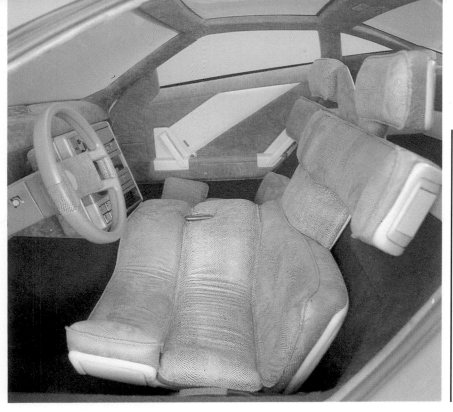

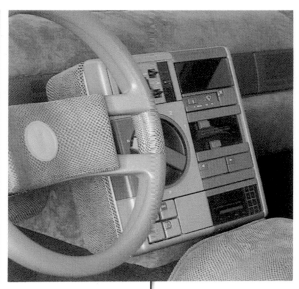

■
Bertone studio's Ramarro, from drawing board concept to running prototype.

MARCELLO GANDINI

The son of a professional musician, Marcello Gandini was born in Turin in 1939, and after dropping out of studying classics at school, he worked for a body repair shop. His first creation was in 1959 on an OSCA Barchetta, from which he learned a great deal about vehicle construction methods. His first break came when Giugiaro left Bertone, and he found himself hired as chief designer. Gandini's first job here was to get some good designs ready for the Geneva show, just four months away. The spectacular Lamborghini Miura was the result, and Gandini's reputation in the automotive business soared.

Many commissions came to Bertone afterwards, including the Lamborghini Marzal and 4-seater Espada, Ferrari's one and only Bertone body on the 308GT4, the dart-shaped Maserati Khamsin, the Alfa Romeo 1750 Berlina, Innocenti's Mini, the Fiat X-1/9, and the rally championship-winning Lancia Stratos, all of which had extensive if not total Gandini input. Most dramatic of all was the Lamborghini Countach of 1971, which with certain updating is still as radical a design in 1989 as it was 18 years ago. His last job at Bertone was the styling for the Citroen BX saloon, which the French manufacturer wanted to appeal to a wider motoring public than most of their previous models had done.

Gandini left Bertone in 1978, and today he works from his villa in the wooded hills above Turin. As an independent consultant, his first five years were almost entirely committed to Renault; an early commission was to style, both inside and out, the facelifted R4; next came the interiors for both R25 and Alpine, and he has been asked other proposals for Renault's truck division. In 1988, he styled the dramatic V16-engined Cizeta Moroder for a consortium of ex-Lamborghini employees based in Modena. Back in the supercar league again, Gandini was responsible for the Lamborghini Diablo, a perfectly suitable successor to the Countach, but certainly not such an innovative design. In the midst of Fiat's proposed takeover of Maserati, his latest creation for the latter was the successor to the Bora-Merak, a 4.9-litre mid-engined V8. Like Giugiaro, Gandini has also recently expanded his design consultancy horizons to take in clocks, lamps and bicycles.

TOP Citroen BX 1.9 GTi at the 1986 Motor Show.
CENTRE AND RIGHT Gandini generally gets the credit for the Lamborghini Muira, styled when he was at Bertone.

The Lamborghini Muira symbolised the '60s supercar, whilst the Countach was its '80s equivalent.
CENTRE AND LEFT Lancia Stratos HF, World Championship-winning rally car.

ALEC ISSIGONIS

Sir Alec Issigonis began drawings for the Morris Minor during the Second World War. The car was produced in several forms from 1948–71.

Alec Issigonis was born in Izmir in 1906, into a family involved in marine engineering, and as a result of political machinations after World War I when the business was confiscated by Turkey, Alec came to to Britain to study engineering at Battersea Polytechnic. After a European tour with his mother, he found a job as chief draughtsman at Edward Gillett's in London. Gillett was developing an automatic clutch – inferior to the synchromesh gearbox then being evolved by General Motors – but the work introduced Issigonis to a number of British manufacturers, and he was soon offered a job at Humber. This posting was swiftly followed by another, to Morris Motors at Cowley. At this time a centralized engineering division was being set up at Cowley, with separate cells for each part of the vehicle; Alec was responsible for the back axle! He was soon designing the whole of the suspension system.

As a hobby, Alec began racing a 'Lightweight Special', based on his Austin 7 Ulster, with George Dowson, and it was while competing in sprints and hillclimbs that he learned a lot about suspension and body-stiffness, as well as general driveability and handling. But for the war, his coil-spring and wishbone independent suspension system, together with his rack and pinion steering, would have debuted on the Morris series M10. These components were actually first used on the 1947 MG Y-Type in 1947.

All through the war, he had been making up designs for what would become the Morris

Minor, and was given a free hand by chief engineer Vic Oak. He was assisted by two draughtsmen, Jack Daniels on chassis, suspension and steering, and Reg Job on the body. The car was launched to much acclaim at the 1948 London Motor Show: except that at the trade launch, Lord Nuffield snubbed it as looking like a 'poached egg'!

When the Nuffield Organisation, which owned Morris, merged with Austin to form the British Motor Corporation in 1952, Issigonis left to join Alvis, assuming, probably quite rightly, that his Morris designs would be temporarily shelved. Unfortunately, his first project at Alvis was abandoned; Alvis decided that too much development had gone into the car to make it profitable, including a 3.5-litre 90-degree V8 engine, plus the brand-new Moulton Hydrolastic suspension system, so it was shelved. By 1956, Issigonis was back at BMC under the chairmanship of Sir Leonard Lord.

At this point the Suez crisis intervened; the Suez canal was blocked by Egypt, and the oil pipeline through Syria was cut. The whole of Europe was forced to reconsider its perception of the automobile, one of the swiftest reactions being the sudden abundance of economical German and Italian bubble cars. As far as Lord was concerned, this was intolerable, and he encouraged Issigonis to produce something appropriate to combat the influx, using the existing BMC parts bins where possible. Just as he had with the Morris Minor, Alec worked with a multitude of carefully annotated freehand sketches, and between March and July 1957, his small team had built wooden mock-ups. Key engineering innovations were the placement of the gearbox under the engine (in the sump in fact), and the transverse mounting of the engine. Moulton variable rate rubber suspension was used, and the tiny ensemble ran on equally diminutive 10 inch Dunlop wheels. A year later, most of the bugs had been ironed out, and by mid 1959 the Mini was in production. This was Issigonis' second complete design. The Mini was followed by a line of similarly conceived BMC family cars, increasing progressively in size, but none achieved as much as the Mini; it was for a time unbeatable in competition, and it revolutionized the European motor industry. Alec Issigonis received many accolades, including a knighthood in 1967. He died in 1989.

Issigonis is best remembered for the front-wheel-drive, transverse-engined Mini of 1959, which virtually revolutionised the European motor industry. Thirty years on it is still in production.

MALCOLM SAYER

Malcolm Sayer learned his skills as an aero-dynamicist where they really matter, in the air-craft industry. His first job at Sir William Lyons' Jaguar Cars was the body styling for the com-pany's first purpose-built racing car, the C-Type. Remarkable for its purity of line, the C-Type was an instant success, being driven to victory in the 1951 Le Mans 24-Hour race by Peters Walker and Whitehead. In the same event in 1953 the C-Types finished a remarkable first, second and fourth. Sayer's next project was the D-Type, on which the curves of the wings and wheelarches were more accentuated, almost bulbous. There was an open air intake instead of the C-Type's grille, and behind the driver's cockpit was the characteristic headrest and tail-fin, intended to promote high-speed stability. The D-Type was one of the most successful rac-ing cars of the mid-to late 1950s.

Malcolm Sayer's other creation was the leg-endary E-Type, the only road-going Jaguar which was not the product of Sir William Lyons. Before his premature death, Sayer had styled the mid-engined Jaguar XJ-13, a long, low, rounded sports prototype which was tested, crashed, and rebuilt as a museum piece in 1967.

BELOW Purest of Malcolm Sayer's creations was the Jaguar C-Type, Le Mans winner in 1951.

BELOW RIGHT The prototype D-Type Jaguar of 1954.

Last of the works D-Types, chassis XKD 406, campaigned in 1958.

■ **OPPOSITE PAGE, TOP AND LEFT** Sayer's last opus was the mid-engined Jaguar XJ-13 sports-racing prototype built in 1967.
BELOW Phallic symbol of the 1960s, the E-Type roadster.

WILLIAM TOWNS

Born in 1936, William Towns joined Rootes Group in 1955, and was delighted to find that there was a department which did things with plasticine in full scale in exactly the same way as he had in miniature as a child. During the early 1960s he worked at Rover, designing the development of the gas-turbine Le Mans racing car of 1965. The following year, Towns was offered a post at Aston Martin, and, turning up his nose at the proposed Touring designs for the forthcoming V8 models, he set about drawing up the DB-S. Alongside this he styled the huge, wedge-shaped 4-door Aston Martin Lagonda. Bill Towns also drew replacements for Triumph's TR6, which got lost in the Stokes era at British Leyland, and also did a great deal of the work on the styling of the Jensen-Healey sports car, launched at Geneva in 1972.

RIGHT Bill Towns with the mini-based Minissima town car. Towns' speciality was sporting machines like the 1972 Jensen-Healey and Aston Martin DBS 6 cylinder.

PMH 281L

OGLE DESIGN

Founder David Ogle was killed in a car crash in 1962, but work carried on under John Ogier and stylist Tom Karen. Early projects were a sports car based on a Daimler SP250, and a Mini. The fibreglass-bodied Reliant Scimitar Coupé commission came in 1964, followed four years later by the Scimitar GTE, which set off a new trend in well-appointed sporting estate cars. The GTE is back, courtesy of Middlebridge Engineering. Meanwhile, Ogle did consultancy work for Turkish manufacturer Anadol, going on to design the antithesis of the Scimitar, that three-wheel wedge, the Bond Bug. Continuing on the three-wheeler, fibreglass body theme, the Bug fun-car was replaced at Reliant by the dull but practical Robin. At the other end of the vehicular spectrum again, Ogle styled an Aston Martin V8 for the 1972 Montreal show. Only two cars were built, but the Ogle Aston went some way to proving that styling houses other than those of Detroit or Italy had panache.

Ogle Design styled this Aston Martin V8 for the 1972 Montreal show.

FRANK COSTIN

From 1955 to 1975, the man at the forefront of automobile aerodynamics was Frank Costin. His designs for racing cars, notably the Lotus Marks 8,9,11, and the road-going Elite, plus the Vanwall and March 711 Formula 1 cars, endowed them with better air penetration than most of their contemporaries. Apart from the March, they carried no artificial aids, and the cars were seemingly unaffected by cross winds. On the race circuit of course, there was no need to compromise with the niceties of fashion, and accordingly his designs, in conjunction with Colin Chapman's engineering skills, often proved faster than cars with similar engine capacities, which is the name of the game.

Costin's other area of expertise was the use of wood as a structural material in car design. Although wood was widely used for framing and panelling in the pioneering days of the motor car, Costin brought his aircraft engineering training at De Havilland to bear in the manufacture of wooden monocoque chassis. His first such car was the 'ugly duckling' Marcos of 1960, conceptually a glider body on wheels, made at

his workshop at Dolgellau, north Wales. Here he had done much development work on Chapman's first 'production' car, the Lotus 7, which itself was beaten on the track by the Marcoses, often driven by great racers of the future such as Jackie Stewart.

A Costin sports racer came out in the mid 1960s, competing against the likes of the Lotus 23, and there was a lightweight wooden shopping car done about the same time. The sports racing cars continued with the beautifully curvaceous Costin-Nathan, which was very successful in GT form during 1967 and 1968, but which eventually proved too expensive to build in quantity. From 1972 to 1980, he produced a Marcos-like GT car called the Costin Amigo. This Vauxhall-powered car could do 130 mph, but was over-long, and extra-low at only 3 foot 6 inches; Costin incorporated a warning light on the radio aerial to warn of the car's approach down narrow lanes. The car was eventually abandoned after a stalemate with backers, and Costin moved to Ireland to take up various kinds of consultancy work.

■ The Lotus Elite was an all-fibreglass coupe, with 1216cc Coventry Climax engine and competition-derived dashboard.

■ The Elite set new handling standards for roadgoing sports cars. About 1000 were built between 1958–64.

KARMANN

Most people associate the coachbuilders Karmann with Volkswagens, and more particularly the Ghia-styled VW coupé of the late-1950s and early 1960s. Karmann goes back a long way though, having built car bodies since the turn of the century. Under its Karmann Engineering branch, it has frequently provided a production line for well-known German models. In addition, it is noted for its machine tool design.

Wilhelm Karmann took over an Osnabruk bicycle manufacturing business in 1901, and began building bodies for Durkopp, Minerva and FN chassis. His first exhibition was at the Berlin show of 1905. The company expanded gradually until World War I, when the 70-strong workforce was cut back to 20, to build Zeppelin winch-cars and ambulances. They were reduced to making agricultural carts for French and Belgian farmers as part of the German reparations after the war. Soon the one-off orders tailed off, and Karmann was building bodies fifty at a time for manufacturers like N.A.G., FN, Protos and AGA. Wilhelm visited the US in 1924 and was impressed with Fisher's practice of spraying steel panels with nitro-cellulose paints. Manufacturers were often going broke in the perilous times of 1920s Germany, and when Mannesman failed to collect their fleet of six-seater bodies, Karmann offered them to his workforce as summer houses to put in their gardens.

The coachbuilder's trade was seasonal, with heavy demand during the spring and summer, and a dearth during the winter. However, Karmann won a substantial contract with Adler following the showing of an award winning cabriolet at the 1926 Bad Nauheim concours.

Other manufacturers turned to Karmann, and from 1936 until 1939, some 800 of the pretty Eifel Cabriolets were built for Ford Cologne. During this pre-war expansion, Karmann's 800 employees built several mundane models for Hanomag, Krupp and Trumpf at a rate of around 65 bodies a day, including commercial vehicles and cabriolets. The factory was the victim of the numerous wartime allied bombing raids, and the restoration of British Humber staff cars saved the Karmann factory from potential closure after the war. From 1949, Karmann produced Cabriolet bodies for Volkswagen's Beetle, and this was followed in 1950 by an order from DKW for 5000 four-seater convertibles, which it produced until 1955. Wilhelm Senior died in 1952, and was succeeded at the head of the company by his son. Wilhelm Jnr. had studied car-body construction and worked as a designer for rivals Ambi-Budd in Berlin, and he had also contributed to the original VW cabriolet project. With the launch of the VW Karmann Ghia in 1955,

ABOVE Volkswagen is one of Karmann's biggest customers. The razor-edge shape of the 1963 Karmann Ghia was made at their Osnabruk factory, as are bodies for the Seat Ibiza.
BOTTOM Both Porsche 356 Coupe and Cabriolet had bodies built by Karmann.

the firm's name was to be seen on a car for the first time. The result of a joint venture between Wilhelm Jnr. and his friend Luigi Segre at Ghia, it took only 14 months to produce 10,000 units. Capacity at Karmann's factory increased dramatically during the late 1950s, and by 1962 they were supplying bodies for the Porsche 356, as well as the 1500cc Karmann Ghia. Naturally, Karmann won the contract to supply bodies for the VW-Porsche joint venture, the mid-engined 914/6 sports car.

By 1966, a second plant was established at Rheine near the Dutch border where bodies for cars such as the big BMW coupés are made, as well as the Ford Merkur XR4Ti for US export. At Osnabruk, meanwhile, Karmann was building bodies for the VW Scirocco coupé and VW Golf cabriolets, and although the majority of its work comes from VW, the firm is currently going from strength to strength with models as diverse as the VW Corrado coupé, the Seat Ibiza, and the Escort convertible.

Karmann builds bodies for models as diverse as the Seat Ibiza, US Ford Merkur XR4i and VW Golf Cabriolet.

Yuppie's delight; the Karmann-made Cabriolet version of the Ford Escort Cabriolet.

HERMANN GRABER

The son of a coachbuilder, Hermann Graber was born in Bern, Switzerland in 1904, and he inherited the family business on his father's death in 1925. He became a car enthusiast while studying in France, but his early work followed on from the coachbuilding heritage: for instance, he made a courtesy coach for the Grand Hotel Mattenhof. The most popular construction material in the 1920s was the Weymann-type fabric-covered plywood, but Graber abandoned it in favour of steel. The first of the new steel bodies were for a 4-litre Delage D6 with cycle mudguards and running boards, and a Voisin with an especially coherent tryptich facade, its rounded grille separating the rounded mudguards. The styling of his Mercedes SSK stressed the brutish purposefulness of this car, but Graber soon found himself working with American chassis which were popular among the wealthy Swiss population, not least because their relative cheapness allowed a greater spending margin for lavish body-styling. One of Graber's best efforts was on a 1937 Duesenburg, given the full Grand Routier treatment, but this was a one-off; his work on US cars was mostly adapting Dodge and Packard saloons and transforming them into sumptuously appointed drophead coupés.

Naturally, Graber also worked on the great chassis of the day, the Delahaye 135 MS, Alfa Romeo 6C 2500, Bugatti and Hotchkiss, and his favourites of the post-war period, the Talbot Lago Record. There was nothing dramatic about these cars; just graceful, gently curving lines bereft of messy running boards, but strictly conservative in character. There was no ornamentation just for the sake of it, and the grille always identified the chassis' origins. Even the original spoked wheels were retained. As the supply of the French chassis diminished, Graber turned his hand to Rover, Armstrong-Siddeley, Bentley, Aston Martin, and finally, Alvis.

This arrangement came about when Alvis lost their regular supplier, Mulliner, to Standard, and although a typically restrained effort, the Graber Alvis of 1955 was warmly received at Earls Court. Despite Graber's illness in 1958, the Alvis TC 108 G was followed by the TD 21 of 1959, with obvious concessions to contemporary US styling in its tail fins and knotch-back rear window. There were Graber Special and Super versions as well. In 1963 came the finely proportioned 4-door Alvis TE 21, of which just four were made, and the Alvis TF 21 naturally followed in 1964. Graber had produced a cabriolet on the Rover P5 body in 1965, and ironically, his major client, Alvis, was taken over by Rover in the same year. Rover's inclusion into the British Leyland conglomerate spelled the end for Alvis, who were perceived as being too closely competitive with Jaguar.

With the demise of Alvis, Graber was without a regular chassis source. As a sort of concession, he was appointed Rover's agent in Switzerland, and made a series of conversions on current Rover models. There was an attractive P6 coupé in 1966, and then a convertible version of the P6, followed by a 3500 coupé in 1968. These conversions were little more than an adjunct to the Rover salesroom, and Graber did no further styling work. He died aged 66 in 1970, and all the paraphernalia of the design studio went up in a giant conflagration at a wake organized by his grieving wife.

Hermann Graber is probably best known for his work on Alvis cars like these 3-litre TFs during the late 1950s and early 60s.

SIXTEN SASON

When Nazi Germany began its Blitzkrieg in Europe in 1939, a radical defence programme was already under way in Sweden. SAAB (Svenska Aeroplan Aktiebolaget) was a company formed to build aircraft as part of this programme, and after the war in 1945, there was sufficient capacity and motivation to consider building a small passenger car. A simple car like the DKW was an obvious choice, producible with the minimum of special tooling. Starting with a clean sheet of paper was difficult, especially since no-one at Saab was a stylist. Sixten

Sason was the technical illustrator who enjoyed sketching, and his initial drawings for the car looked in profile distinctly like a section of an aircraft wing, with a long, sweeping curve for the roof-line. Logical really, in an aircraft manufacturing company. The wheels and headlights were fared-in by the sweeping bodywork, although these features were rationalized for the production car.

Born in 1912, and an art student in Paris, Sason (a contraction of the name Andersson) lost a lung in a flying accident, but after recovering turned his hand to designing a variety of objects, from sewing machines for Huskvarna to delta-wing jet fighter planes. One of his drawings of the 1930s pre-dated the Mini's transverse engine and gearbox by 25 years. By 1955 he had designed parts for the Hasselblad camera system later to be used by US astronauts.

Sixten's brief for the Saab car, known as project 92, was fairly open. The car needed to be rugged to withstand the rigours of poor roads and Sweden's harsh winter climate. It was decided that the engine should be mounted at the front, in order to keep most of the mechanical parts in one place, and to maximize the area of the passenger compartment. The advantages in road behaviour of front-wheel-drive in this type of car had not been considered, but an efficient independent suspension system of torsion bars and telescopic dampers no doubt contributed to the overall efficiency of the design. The chief problems in developing the prototype Saab were finding suppliers to co-operate and other models to dissect for evaluation and comparison. Nonetheless, a prototype was finished in 1946, and it was found in wind-tunnel tests that

Sixten Sason and prototypes of the first Saab 92. Sweeping lines show aircraft design influence.

Sason's aerodynamic shape endowed it with an astonishingly low drag coefficient of 0.32. At the time, perhaps only the French makes Citroen and Panhard were as equally concerned with aerodynamic efficiency. The car was of unit-construction, and its floor-pan completely enclosed its underside. It was powered by a 764cc vertical two-stroke engine, producing just 25hp, but nevertheless, it achieved a reputation for being just the car for obtaining a high average speed over rough and winding roads, particularly when they were covered in ice and snow.

Further pre-production problems arose: firstly in acquiring body presses, which were eventually sourced in Chicago, and then in delivering and installing them at the Trollhattan plant. But by 1949, the Saab 92 was in production. How good its original conception was is shown by the astonishing longevity of its production run, which lasted, with suitable updating and increases in engine capacity, until 1980, and by the innumerable rallying successes, including Eric Carlsson's Monte Carlo wins in 1962 and '63.

1949 prototype of the Saab 92. 001 is in Saab's museum at Trollhattan.
BOTTOM Sason's drawing of the 2-stroke twin-cylinder engine of 1954.
BOTTOM RIGHT Enthusiast-converted Saab 96 Cabriolet.

After the 92 series cars, Sason produced ideas for the subsequent 99 range, one of which can be seen in the Saab styling department, left.

INDEX

Thanks to the following organizations and persons for their help with this book:
Chuck Jordan, Floyd Joliet and Richard Stoey at General Motors, USA; Chuck Ordowski at Ford USA; Kyle Johnson and Jim Falk at Cadillac USA; Nancy Cox at Chrysler Historical Collection USA; Oldsmobile, USA; Linda Giglio at the MVMA, Detroit, USA; Buick Motors, USA; The Henry Ford Museum, Deerborne, USA; Fiat Auto (UK) Ltd; Fiat Centro Storico, Italy; Centro di docmentazione Alfa Romeo, Italy; Archivo Photografico Lancia, Italy; Maurizin Ilrhana at Italdesign, Italy; Mr Pagnbi at Pininfarino, Italy; Ferrari Spa, Italy; Andrew Morland Collection; National Motor Museum; Quadrant Picture Library, Sutton, Surrey, England